Rethinking and Relearning Disaster Adaptations from and within Indigenous Land-Based Perspectives

This book offers a critical exploration into Indigenous knowledge systems, particularly focusing on Indigenous land-based knowledge and practice in reshaping disaster adaptations.

Drawing from Indigenous communities in Bangladesh, this book challenges transformational approaches to disaster resilience by centering on land-based perspectives intrinsic to Indigenous cultures. The book showcases how Indigenous and land-based minority communities in Bangladesh have historically coped with and adapted to environmental challenges. It navigates beyond the Eurocentric paradigm, acknowledging the richness of traditional Indigenous land-based knowledge and practice embedded in the relationship between Indigenous peoples, land-based minority communities, and their natural environments. The book focuses on the interconnectedness of Indigenous land-based knowledge, culture, and sustainable practices, providing a blueprint for rethinking contemporary disaster adaptation strategies. By relearning from Indigenous land-based perspectives, readers gain invaluable insights into holistic, community-based approaches prioritizing harmony with nature over technological fixes. Through Indigenist, decolonial, relational, and feminist theoretical research frameworks, the book advocates for a paradigm shift in disaster management, emphasizing the importance of respecting and integrating Indigenous land-based solutions.

Rethinking and Relearning Disaster Adaptations from and within Indigenous Land-Based Perspectives emerges as a crucial resource for scholars, policymakers, and practitioners seeking to foster resilience through a more inclusive and culturally sensitive lens.

Ranjan Datta is Canada Research Chair in Community Disaster Research at Indigenous Studies, Department of Humanities, Mount Royal University, Calgary. Alberta, Canada. Dr. Datta's current research interests include advocating for Indigenous land-rights, Indigenous community disaster research, community resiliency on climate change, community-based participatory action research, decolonization, and Indigenous reconciliation. Dr. Datta published 97 peer-reviewed publications, four books, and six edited books on responsibilities on decolonization, cross-cultural perspectives on reconciliation,

Indigenous water, Indigenous climate change, anti-racist perspectives on climate change, and environmental sustainability issues. Dr. Datta has developed a strong understanding of decolonial and Indigenist research frameworks in his 17 years conducting research with Indigenous and non-Indigenous communities in Canada, USA, Africa, Europe, and South Asia. He is strongly committed to and passionate about Indigenous environmental sustainability, reconciliation, environmental management, Indigenous land rights, anti-racist theory and practice, decolonization, social and environmental justice, community gardens, and cross-cultural research methodology and methods.

Margot Hurlbert is a Professor and Canada Research Chair, Tier 1, Climate Change, Energy, and Sustainability Policy of the Johnson Shoyama Graduate School of Public Policy, University of Regina. She explores the gap between what is needed to address climate change and current policy and behavior. Margot's scholarship concerns climate change adaptation and mitigation, energy, Indigenous peoples, water, droughts, floods, water governance, and sustainability. Margot has authored IPCC AR6 and Special Reports and was a member of Future Earth's Earth Commission Working Group on Transformations (2019–2022). She is an expert panel member on "Adaptation" for the Canadian Climate Institute and an Expert with the Canadian Climate Law Initiative.

Arifatul Kibria is a feminist scholar who actively supports and promotes the rights and interests of marginalized and disadvantaged communities. She aims to amplify the voices of Indigenous and minority communities through an educational framework that incorporates decolonial and Indigenist perspectives. She holds the position of Associate Professor in the Department of Social Sciences within the Faculty of Arts and Social Sciences at American International University-Bangladesh. She has a diverse range of academic interests, which encompass sociological theories, gender studies, governance, social change and mobility, migration, and climate change. In 2002 and 2001, respectively, she obtained her Master of Social Sciences (MSS) and Bachelor of Social Sciences (BSS) degrees in sociology from the University of Dhaka. In 2011, she earned her PhD in social structure, stratification, and inequality from the Department of Sociology at Renmin University of China.

Rajmoni Singha is an Assistant Professor at the Department of Sociology, at North South University, Bangladesh. Dr. Singha earned his PhD in anthropological research focusing on Manipuri weavers and their challenges in Bangladesh at Charles Darwin University in Australia and Master's degree in Master of Development Practice(MDP) from James Cook University, Australia. Singha's broad research areas are anthropological research, contemporary development challenges, qualitative research methods in social sciences, health and environment, climate change and health, Indigenous knowledge and environmental management, and ethnicity and identity studies.

Barsha Kairy is a Master's Student at the Johnson Shoyama Graduate School of Public Policy, University of Regina. She holds a Bachelor's and Master's in Education from the University of Dhaka and has worked with various organizations. She has a passion for Social Science and has pursued a second Master's in Development Studies. She currently works in the development sector, focusing on research on Indigenous communities, gender, disaster resilience, climate change, gender, and public policy. Her goal is to dedicate herself to academia for better service.

Somashree Chattapadhya is a Master's Student at the Natural Resource Institute (NRI) at the University of Manitoba, Canada. Her research focuses on community empowerment through addressing climate change, water and sanitation, agriculture, etc. She is passionate about working with Indigenous and minority communities as she believes in changing the community through her work. She works for Water.org as a Program Manager, providing household-level solutions for accessing water and sanitation, especially in climate-vulnerable areas. She is interested in exploring the gender aspect of climate change, water resource management, Indigenous land-based studies, etc. She completed her master's degree in environmental economics as a second topper from the University of Dhaka in 2015.

Routledge Explorations in Environmental Studies

Myth and Environmentalism
Arts of Resilience for a Damaged Planet
Edited by Esther Sánchez-Pardo and María Porras Sánchez

Christian Environmentalism and Human Responsibility in the 21st Century
Questions of Stewardship and Accountability
Edited by Katherine M. Quinsey

Frontier Thinking and Human-Nature Relations
We Were Never Western
E. C. H. Keskitalo

Ecophenomenology and the Environmental Crisis in the Sundarbans
Towards a Community-Based Ethic
Kalpita Bhar Paul

Indigenous Land-Based Knowledge and Sustainability
Settler Colonialism and the Environmental Crisis
Edited by Ranjan Datta, Jebunnessa Chapola and John Bosco Acharibasam

Intergenerational Democracy, Environmental Justice and the Case of Nuclear Waste
Lee Towers and Matthew Cotton

The Energy Trilemma in the Baltic Sea Region
Security, Equity and the Environment
Edited by Michael Kalis

For more information about this series, please visit: www.routledge.com/Routledge-Explorations-in-Environmental-Studies/book-series/REES

Rethinking and Relearning Disaster Adaptations from and within Indigenous Land-Based Perspectives

Ranjan Datta, Margot Hurlbert, Arifatul Kibria, Rajmoni Singha, Barsha Kairy and Somashree Chattapadhya

Routledge
Taylor & Francis Group
LONDON AND NEW YORK

earthscan
from Routledge

First published 2025
by Routledge
4 Park Square, Milton Park, Abingdon, Oxon OX14 4RN

and by Routledge
605 Third Avenue, New York, NY 10158

Routledge is an imprint of the Taylor & Francis Group, an informa business

British Library Cataloguing-in-Publication Data
A catalogue record for this book is available from the British Library

ISBN: 978-1-032-88133-1 (hbk)
ISBN: 978-1-032-88134-8 (pbk)
ISBN: 978-1-003-53634-5 (ebk)

DOI: 10.4324/9781003536345

Typeset in Times New Roman
by SPi Technologies India Pvt Ltd (Straive)

Contents

1 Reclaiming Indigenous Community-Led Disaster Adaptation

Decolonizing human-created environmental disasters and relearning adaptations from Indigenous land-based perspectives in Bangladesh explore how Indigenous communities' social, economic, and environmental considerations aligned with cultural and traditional values. Resilience to disasters within Indigenous contexts could not be comprehended in isolation from the community's land-based traditions, land rights, and customary practices. While the pursuit of disaster crisis resilience was broadly conceptualized, it was deeply rooted in traditional land-based learning and practices, as land-based knowledge held hybrid, relational, and scientific meanings for Indigenous communities. This book systematically examines how Indigenous communities revitalized the meanings of disaster adaptations by leveraging traditional knowledge to seamlessly integrate traditional sentiments with the large-scale dislocations occurring both within their land-based knowledge and practice. Using Indigenist, decolonial, relational, and feminist theoretical research frameworks, the book addresses fundamental questions surrounding disaster adaptations concerning conceptions and practices of land management. It seeks to explore effective ways those invoking the term "disaster adaptations" could proactively engage with Indigenous ecological, economic, and social challenges. In line with the specified research questions, the book was guided by critical concerns in identifying the problems inherent in existing forest and disaster management practices concerning everyday land-based practices and traditional experiences in Indigenous regions.

Benefits of This Book

A growing cohort of educators, social and climate change practitioners, and researchers actively sought effective, ethical, and appropriate ways to incorporate Indigenous land-based knowledge systems(ILBKS) into climate resiliency initiatives. However, colonial history often led to the misrepresentation or misunderstanding of Indigenous people, ILBKS, and their land-based culture globally (Battiste, 2017; 2014; Kovach, 2009; Little Bear, 2013). This book aims to foster solidarity with Indigenous peoples' struggle for meaningful land-based disaster adaptation, celebrate diversity, and engage in intimate, cross-cultural relations with Indigenous communities. It draws on various perspectives

DOI: 10.4324/9781003536345-1

and paradigms to offer substantial benefits to Indigenous and non-Indigenous communities, particularly those educated in non-Indigenous ways or lacking exposure to ILBKS. Addressing epistemological and cosmological differences, the study explores how to integrate traditional Indigenous land-water customs and practices into disaster adaptation strategies. It emphasizes the recognition and respect of ILBKS, honoring Indigenous perspectives for more sustainable ways of being in interdisciplinary communities involving both Indigenous and non-Indigenous individuals working toward disaster adaptation cooperation with cross-cultural knowledge, limits, and possibilities.

The book critically explores the global significance of disaster adaptations, especially from Indigenous land-based perspectives on disaster adaptations. It acknowledges the positive impacts of traditional land-based perspectives on disasters witnessed in Australia, Canada, New Zealand, and other international Indigenous contexts. ILBKS's perspectives on disaster adaptation practices were pertinent to academic specialists in disaster research, disaster ministries, and policymakers across diverse regions, including Australia, Canada, New Zealand, South Asia, Norway, and South and North America. Despite the growing field of disaster resiliency, Indigenous agendas often did not align, particularly in South Asia.

The research benefits a wide audience, including Indigenous and non-Indigenous educators, students, faculty, researchers, activists, and environmental professionals involved in disaster resiliency and adaptation policies. It holds particular value for diverse communities, streamlining the process of imparting basic understandings to Western-trained educators, professionals, immigrants, and refugee citizens. Although its initial focus was on Bangladesh's Indigenous communities, the research anticipates high relevance to those working in South Asia and other communities across Asia and beyond.

The book caters to critical readers in disaster research, environmental sociology, anthropology, disaster research, interdisciplinary studies, postcolonial studies, ethnic studies, environmental sustainability, and Indigenous and women's studies. Its interdisciplinary nature draws upon diverse fields such as postcolonial theories, Indigenous methodologies, sustainability theory and practices, decolonization, community-based practice, youth practice, and Indigenous knowledge and practice, ensuring broad appeal to academics and practitioners across various disciplines. Embracing a "decolonization and reclaiming approach," the book also addresses the educational policymaking community in the Eastern and Western worlds. It takes a significant step in implementing relational interdisciplinary meanings of sustainability, emphasizing responsibilities for building reconciliation in education (Battiste, 2017, 2014). The research contributes to various intersections between critical anti-racist education and decolonization, impacting different aspects of Indigenous and non-Indigenous lives. It explores holistic meanings of people's lives and their responsibilities for Indigenous education and environmental sustainability, protecting traditional knowledge, culture, and community resilience. Discourses and practices surrounding responsibilities for climate resiliency, local outcomes of cross-cultural bridge, cultural appropriate knowledge, social and environmental justice, and solidarity are of special interest.

Beyond these academic contributions, the book has multiple benefits, such as finding meanings for both the research and researcher, centering local Indigenous perspectives on disaster, and challenging colonization that persisted in the aftermath of colonialism, ongoing racism, and the exclusion and Othering of Indigenous knowledge in their own land.

Book Context

Indigenous communities in Bangladesh, particularly those dwelling in coastal and hilly areas, confront severe climate challenges due to geographical, ecological, and socioeconomic factors. However, many climate crises in Indigenous communities stem from enduring ongoing settler colonialism, external encroachments involving land grabbing, colonial farming practices, and external environmental governance within Indigenous territories (Datta, 2023, 2019, 2017; Chakma, 2010a; Roy, 2010). In the coastal and hilly areas, Indigenous communities face unique climate challenges stemming from their reliance on forest ecosystems and traditional agricultural practices (Garai et al., 2022; Rasul, 2007; Ullah et al., 2022). Forest deforestation, often linked to external development projects, reduces the natural buffer against landslides and alters precipitation patterns, affecting water availability and soil stability (Rasul, 2007; Ullah et al., 2022). These disasters disrupt age-old practices that have allowed Indigenous communities to coexist harmoniously with their surroundings, pushing them into a precarious balance between preserving their cultural heritage and adapting to the evolving climate (Chakma et al., 2020).

Historical injustices also play a significant role in increasing the human-created environmental disaster challenges faced by Indigenous communities in Bangladesh (Adnan, 2010; Baird, 2008; Datta & Chapola, 2018; Roy, 2000a). Encroachment on Indigenous lands, often driven by settler colonialism and external interests, exacerbates the vulnerability of these populations (Mason & Rigg, 2019). Displacement due to land grabbing and infrastructural projects disrupts established ways of life and further marginalizes Indigenous communities, leaving them at the mercy of increasingly unpredictable and extreme climatic conditions (Datta & Marion, 2021). The intersection of these factors creates a complex web of challenges for Indigenous communities in coastal and hilly areas, necessitating a nuanced understanding and targeted interventions to ensure their resilience in the face of a changing climate.

In Bangladesh, Indigenous communities, deeply connected to their ancestral lands, face a unique set of challenges exacerbated by the interplay of climate change and colonial legacies (Chakma & Sultana, 2023). This book explores the imperative of decolonizing the discourse on climate disasters and relearning adaptations through the lens of Indigenous land-based perspectives in Bangladesh. By critically exploring the historical injustices, acknowledging the unique wisdom embedded in Indigenous knowledge systems, and advocating for a paradigm shift in climate policies, this book aims to uncover a path toward sustainable and equitable resilience for these marginalized communities (Barkat et al., 2010; Johnson et al., 2022; Das, 2011; Van Schendel, 2023).

Understanding the contemporary human-created environmental disaster challenges faced by Indigenous communities in Bangladesh necessitates an exploration of historical legacies, notably the impacts of colonization (Johnson et al., 2022; Das, 2011; van Schendel, 2023). The imposition of external systems of governance and economic exploitation has led to land dispossession, cultural erosion, and socioeconomic marginalization. These historical injustices have created a vulnerable context for Indigenous communities, leaving them susceptible to the compounded impacts of climate change.

Indigenous communities in Bangladesh have cultivated an intimate and symbiotic relationship with their land over generations (Adnan, 2010; Roy, 2000a). Their unique land-based perspectives encompass a profound understanding of ecosystems, weather patterns, and sustainable resource management. By decolonizing disaster discourse, this book advocates Indigenous land rights as the center of adaptation to the climate crisis (Hoque et al., 2019). Indigenous knowledge systems offer a rich repository of adaptive strategies and a holistic worldview that integrates environmental stewardship, cultural identity, and community resilience.

Bangladesh is particularly susceptible to disaster due to its geographical location, low-lying topography, and dense population (Mallick, 2019). Rising sea levels, increased cyclonic activity, and altered precipitation patterns exacerbate the vulnerability of Indigenous communities, particularly those in coastal areas and the Chittagong Hill Tracts (CHT) areas. These challenges intersect with the historical impacts of settler colonialism, creating a complex web of adversities for Indigenous populations.

Decolonizing disasters involves acknowledging and rectifying the historical injustices that contribute to the heightened vulnerabilities of Indigenous communities in Bangladesh (Datta, 2024a, 2024b, 2023). This requires dismantling systems of oppression, ensuring land restitution, and empowering Indigenous voices in decision-making processes. Reframing disasters within a decolonial framework can address the root causes and foster a more just and equitable approach to climate adaptation (Chakma & Maitrot, 2016). Relearning adaptations involve shifting from mainstream governance and technocentric approaches to embracing Indigenous-led, community-driven solutions. Indigenous communities in Bangladesh have developed adaptive strategies that draw on their deep understanding of local ecosystems (Rasul, 2007). These may include sustainable traditional agriculture practices, community-based early warning systems, and traditional water management techniques (Datta, 2019). Relearning these adaptations enhances Indigenous communities' resilience and contributes to the broader discourse on sustainable and culturally grounded climate solutions.

Recognizing and protecting Indigenous land rights is central to decolonization and relearning adaptations (Datta, 2019; Roy, 2000a). Land rights are a matter of justice and fundamental to climate and disaster resilience. Secure land rights allow Indigenous communities to continue practicing sustainable land management, preserving biodiversity, and maintaining the delicate balance between human activities and the environment. Upholding Indigenous land rights is crucial to ensuring the long-term viability of adaptation strategies grounded in Indigenous wisdom.

A crucial aspect of decolonizing disasters is the adoption of Indigenous communities' traditional land-based approaches in Bangladesh (Dewan, 2021; Muthukrishnan & Datta, 2023). Indigenous communities should be leaders in the decision-making processes that affect them. This involves recognizing and respecting diverse knowledge systems, fostering collaborative research, and promoting the agency of Indigenous individuals in shaping their destinies. Community-centric approaches empower Indigenous communities to adapt and innovate in ways that align with their cultural values and sustainable practices.

Decolonizing disasters and relearning adaptations from Indigenous land-based perspectives in Bangladesh are ethical and practical necessities in a rapidly changing climate (Datta, 2019). By acknowledging historical injustices, embracing Indigenous wisdom, and prioritizing community-centric approaches, we can pave the way for more equitable and sustainable climate solutions (Demos, 2020). Upholding Indigenous land rights and integrating traditional knowledge into climate policies benefits Indigenous communities and enriches the broader discourse on resilience and adaptation in the global fight against climate change. As we strive for a more just and sustainable future, we must recognize the interconnectedness of environmental justice, cultural heritage, and the collective well-being of all communities, particularly those on the frontlines of climate vulnerability.

Historical Background of Disasters in Bangladesh

Bangladesh, situated in the delta of the Ganges, Brahmaputra, and Meghna rivers, is one of the most disaster-prone countries in the world. Its geographic location, combined with high population density and socioeconomic challenges, makes it particularly vulnerable to natural hazards. This vulnerability is exacerbated for Indigenous communities, who often reside in marginalized areas and rely heavily on natural resources for their livelihoods. This section provides a brief history of human-created environmental disasters in Bangladesh and their impacts on Indigenous communities, highlighting key events and the role of traditional knowledge in disaster resilience.

Major Human-created Environmental Disasters in Bangladesh

The 1970 Bhola Cyclone: The Bhola cyclone, which occurred on November 13, 1970, is one of the deadliest tropical cyclones on record. It struck the coastal regions of Bangladesh, then East Pakistan, killing an estimated 300,000 to 500,000 people (Chowdhury et al., 2021). The cyclone caused extensive destruction to housing, agriculture, and infrastructure, leaving millions homeless. The Indigenous communities in the coastal areas were severely impacted, losing not only their homes but also their traditional fishing and farming livelihoods.

The 1998 Floods: The 1998 floods were among the most devastating in Bangladesh's history, submerging two-thirds of the country for over two months (Hossain & Sakai, 2021). These floods affected 30 million people, causing widespread damage to crops, homes, and infrastructure. Indigenous communities in flood-prone areas, such as the CHT, experienced severe disruptions. Traditional knowledge, such as elevated homesteads and floating vegetable gardens, played a crucial role in mitigating some of the impacts (Haque et al., 2021a).

The 2007 Cyclone Sidr: Cyclone Sidr made landfall on November 15, 2007, causing extensive damage across southern Bangladesh. The cyclone resulted in over 3,000 deaths and significant destruction to homes, crops, and infrastructure (Paul & Routray, 2022). Indigenous communities in the Sundarbans, a UNESCO World Heritage Site, were particularly affected. The cyclone disrupted their traditional livelihoods, such as fishing and honey collection. However, community-based early warning systems and cyclone shelters, informed by local knowledge, helped reduce casualties (Ahmed & Alam, 2021).

The 2009 Cyclone Aila: Cyclone Aila struck on May 25, 2009, causing widespread damage, especially in the southwestern coastal regions. The cyclone led to severe flooding and salinity intrusion, affecting over 3 million people (Nasher et al., 2022). Indigenous communities, dependent on agriculture and fisheries, faced significant challenges in recovering from the environmental changes brought by Aila. Traditional practices, such as rainwater harvesting and Indigenous crop varieties, were vital in the community's adaptation efforts.

The 2017 Landslides: In June 2017, heavy monsoon rains triggered landslides in the CHT, resulting in over 150 deaths and significant property damage (Islam et al., 2022). Indigenous communities in the hill tracts, including the Chakma, Marma, and Tripuri, were particularly vulnerable due to their settlements on steep slopes. Traditional land-use practices and community solidarity played crucial roles in rescue and recovery efforts.

The 2023 Flood: In 2023, the CHT and wetland areas of Bangladesh experienced severe flooding, impacting Indigenous communities who depend on the region's river systems for their livelihoods (Datta & Kairy, 2024). The unprecedented floods caused extensive damage to homes, agricultural lands, and infrastructure, displacing many families. Traditional knowledge and practices, while valuable, struggled to cope with the scale of the disaster. The increased frequency and intensity of such events underscore the urgent need for comprehensive disaster management strategies that incorporate Indigenous insights and enhance community resilience.

Impacts on Indigenous Communities

Indigenous and minority communities in Bangladesh are disproportionately affected by human-created disasters due to their socioeconomic marginalization and dependence on natural resources. These communities often reside in remote and environmentally vulnerable areas, such as coastal regions, floodplains, and hill tracts, which expose them to higher risks during natural disasters (Rahman & Khan, 2022).

Loss of Livelihoods: Natural disasters frequently disrupt the livelihoods of Indigenous communities. Cyclones and floods can destroy fishing equipment, livestock, crops, and natural resources essential for subsistence. For example, the Mro and Khumi communities in the CHT rely on *jhum* (shifting) cultivation, which is highly susceptible to landslides and heavy rains (Chakraborty et al., 2023).

Displacement: Many Indigenous communities face displacement due to recurring natural disasters. The loss of homesteads and arable land forces them to migrate to safer areas, often leading to cultural disintegration and loss of traditional knowledge. The displacement also creates tensions with host communities, as seen with the land disputes involving the Indigenous peoples of the CHT (Hossain & Deb, 2021).

Health Impacts: Disasters exacerbate health issues among Indigenous populations, who already have limited access to healthcare. Floods and cyclones often lead to outbreaks of waterborne diseases, malnutrition, and mental health problems. The disruption of healthcare services and lack of clean water and sanitation facilities further complicate recovery efforts (Alam et al., 2022).

Cultural Erosion: The frequent displacement and disruption of traditional livelihoods due to natural disasters can lead to the erosion of cultural practices and languages. Indigenous knowledge, which is often transmitted orally through generations, is at risk of being lost when communities are fragmented or relocated (Rahman & Biswas, 2021).

Role of Indigenous Land-Based Knowledge in Disaster Resilience

Despite their vulnerabilities, Indigenous communities in Bangladesh have developed and maintained rich traditions of disaster resilience. These traditional land-based practices, passed down through generations, are crucial for adapting to and mitigating the impacts of natural hazards.

Elevated Homesteads: Many Indigenous communities construct their homes on stilts or elevated platforms to protect against floods and tidal surges. This practice, seen among the Garo and Santal communities, helps reduce damage during floods (Islam et al., 2022).

Early Warning Systems: Indigenous knowledge systems often include sophisticated early warning signals based on environmental cues. For example, the fisherfolk communities in the Sundarbans use changes in wind patterns, animal behavior, and tidal rhythms to predict approaching storms (Haque et al., 2021b).

Sustainable Agriculture: Indigenous agricultural practices, such as the use of flood-resistant crop varieties and floating gardens (*baira*), help communities cope with extreme weather conditions. These techniques ensure food security during and after disasters (Ahmed & Alam, 2021).

Community Solidarity: Strong social networks and community cohesion are hallmarks of Indigenous societies. During disasters, community members collaborate in rescue and relief operations, share resources, and support each other in recovery efforts. This solidarity is crucial for survival and rebuilding (Rahman & Khan, 2022).

Adaptive Resource Management: Indigenous communities have developed adaptive management practices for natural resources, such as sustainable fishing, forest conservation, and water management. These practices enhance ecosystem resilience and reduce vulnerability to natural hazards (Karim & Mimura, 2021).

The historical background of disasters in Bangladesh highlights the country's vulnerability to various hazards, with significant impacts on Indigenous communities. Major events like the 1970 Bhola cyclone, the 1998 floods, and cyclones Sidr and Aila have caused extensive damage and disruption. However, Indigenous knowledge and traditional practices have played vital roles in enhancing resilience and adaptation. By integrating Indigenous knowledge with modern disaster risk reduction strategies, Bangladesh can develop more effective and sustainable approaches to managing natural hazards.

Indigenous Land-Based Knowledge as Sustainable Disaster Adaptation

In this book, we refer to the Indigenous land-based knowledge systems (ILBKS) as the understandings of land-based skills, and philosophies developed by societies with long histories of interaction with their natural surroundings. These systems encompass a wide range of expertise in areas such as traditional agriculture, traditional resource management, health, and weather prediction. The ILBKS is crucial for sustainable development and environmental conservation, as it embodies practices often more attuned to local ecosystems than modern scientific approaches (Berkes, 2018).

Significance in Disaster Adaptation

Indigenous knowledge is vital in disaster adaptation, particularly in regions frequently affected by environmental hazards. These systems offer time-tested strategies for predicting, preparing for, and recovering from disasters. The knowledge is deeply rooted in the local context, making it highly relevant and practical for the specific environmental challenges faced by Indigenous communities (Hiwasaki et al., 2014).

- *Early Warning Systems*: Indigenous communities often possess keen observational skills and understand environmental cues that indicate imminent natural disasters. For example, changes in animal behavior, wind patterns, and sea conditions are used to predict storms or floods.
- *Sustainable Resource Management*: Traditional practices such as rotational farming, agroforestry, and sustainable fishing help maintain ecosystem balance and reduce vulnerability to disasters. These practices often enhance biodiversity and soil health, which can mitigate the impacts of floods and droughts.
- *Structural Adaptations*: Indigenous architecture often includes features designed to withstand local hazards. Elevated homes in flood-prone areas and stilt houses in coastal regions are examples of structural adaptations that enhance resilience to flooding and cyclones.
- *Community-Based Approaches*: Indigenous communities typically have strong social cohesion and collective action mechanisms. This social structure facilitates efficient and coordinated disaster response and recovery efforts, ensuring that resources and assistance are shared equitably.
- *Knowledge Transmission*: The oral tradition of knowledge transmission ensures that vital survival strategies and adaptive practices are passed down through generations. The continual sharing and refinement of knowledge enhance community resilience over time.

Examples from Bangladesh

Bangladesh is home to diverse Indigenous and land-based minority communities, each with unique knowledge systems that contribute to disaster adaptation. Here are specific examples of Indigenous practices in Bangladesh that enhance resilience to natural hazards.

Chittagong Hill Tracts: The CHTs are inhabited by various Indigenous communities who have developed sophisticated land management and agricultural practices to cope with the region's challenging terrain and climatic conditions (Datta, 2019; Roy, 2024a).

Jhum Cultivation (Shifting Cultivation): This traditional agricultural practice involves clearing a piece of forest land, cultivating it for a few years, and then leaving it fallow to restore its fertility. *Jhum* cultivation helps prevent soil erosion, maintains soil fertility, and supports biodiversity, which is crucial for food security during climatic extremes (Chakma & Sultana, 2021; Datta and Marion, 2021).

Terrace Farming: To mitigate the impacts of heavy rainfall and landslides, Indigenous communities in the CHT practice terrace farming. By creating stepped fields on hill slopes, they reduce soil erosion and water runoff, ensuring more stable agricultural yields (Mukul et al., 2021).

Traditional Bamboo Housing: Bamboo, abundant in the region, is used to construct homes designed to withstand earthquakes and heavy rains. The flexibility and strength of bamboo structures make them resilient to natural hazards common in the hill tracts (Rahman et al., 2020).

Coastal Sundarbans

The Sundarbans, a UNESCO World Heritage site, is home to several Indigenous communities, including the Munda and Mahato, who have adapted to the challenging coastal environment through innovative practices.

Nipa Palm Thatch: Indigenous communities use Nipa palm leaves to thatch their roofs, which are highly resistant to saltwater corrosion. This practice helps protect homes from the saline environment and reduces the need for frequent repairs (Haque & Mondal, 2021).

Floating Vegetable Gardens (Baira): In response to frequent flooding, communities have developed floating gardens. These gardens, constructed from water hyacinth and bamboo, float on water and allow for the cultivation of vegetables during flood seasons, ensuring food security (Islam & Shamsuddoha, 2018).

Mangrove Afforestation: The Indigenous people of the Sundarbans actively participate in mangrove afforestation projects. Mangroves act as natural barriers against storm surges and coastal erosion, providing crucial protection to the coastal inhabitants (Hossain et al., 2016).

Northern Floodplains

Indigenous communities in the northern floodplains, such as the Garo and Santal, employ various strategies to cope with the region's flood-prone environment.

Elevated Homesteads: Homes are built on raised platforms to protect against annual floods. This practice prevents water damage to living spaces and reduces displacement during flood events (Hossain & Sakai, 2021).

Rainwater Harvesting: To address water scarcity during dry periods, Indigenous communities have developed rainwater harvesting systems. These systems collect and store rainwater, providing a reliable water source during droughts and dry spells (Redvers, 2020a).

Indigenous Crop Varieties: The use of traditional crop varieties that are more resilient to flooding and drought conditions helps ensure food security. These crops are often better adapted to local conditions and require fewer inputs than modern hybrid varieties (Chowdhury et al., 2021).

Thus, ILBKS in Bangladesh provides invaluable resources for disaster adaptation. These systems, deeply embedded in the cultural and environmental contexts of Indigenous communities, offer practical and sustainable solutions to the challenges posed by climate change and natural disasters. By integrating Indigenous knowledge with modern scientific approaches, policymakers and practitioners can enhance the resilience of vulnerable communities and ensure more effective disaster risk reduction strategies.

The ILBKS approaches are essential for developing more inclusive, equitable, and effective disaster adaptation strategies. By recognizing and integrating ILBKS, promoting participatory decision-making, and addressing historical injustices, decolonial perspectives enhance the resilience and sustainability of disaster adaptation efforts. Critiquing and transforming conventional approaches that marginalize Indigenous voices and knowledge is crucial for fostering social justice and empowering Indigenous communities in the face of increasing climate-related disasters.

Decolonial Research Frameworks and Methods

In this book, we (as Indigenous and non-Indigenous land-based scholars) have used an interdisciplinary methodological framework that synergizes the Indigenist, relational, decolonial, and land-based research approaches to ensure a holistic and culturally responsive analysis. Our research approaches make us responsible to the Indigenous and land-based minority communities that we had opportunities to learn.

Indigenist Research Approach

The Indigenist research approach is rooted in Indigenous knowledge systems and worldviews, prioritizing the sovereignty and self-determination of Indigenous communities (Smith, 2021). This approach emphasizes the importance of relational accountability, meaning that research must benefit the community and be conducted in a way that honors cultural protocols (Wilson, 2008). Indigenist research also involves reclaiming and revitalizing Indigenous ways of knowing, challenging colonial methodologies (Datta, 2023; Kovach, 2010b). It focuses on the need for research to be reciprocal, community-driven, and respectful of Indigenous spiritual, cultural, and historical contexts (Peltier, 2018; Chilisa, 2020).

Relational Research Approach

The relational research approach centers on the interconnectedness of all beings, asserting that knowledge is co-created through relationships (Wilson, 2008). This approach challenges individualistic and objectivist paradigms by emphasizing the importance of reciprocity, respect, and collaboration in the research process (Hart, 2010). Relational research is grounded in Indigenous and other non-Western epistemologies that value relational accountability and ethical engagement with participants (McGregor, 2018). It recognizes that knowledge is not static but emerges through dynamic interactions among people, places, and the natural world (Tuck et al., 2015; Wilson & Yellow Bird, 2005).

Decolonial Research Approach

The decolonial research approach is concerned with dismantling colonial power dynamics in knowledge production, prioritizing marginalized voices and epistemologies (Smith, 2021). It critiques the dominance of Eurocentric methodologies and advocates for the recognition of multiple ways of knowing (Mignolo & Walsh, 2018). Decolonial research seeks to empower communities by reclaiming and revitalizing Indigenous, Black, and other subaltern knowledge systems (Bhambra, 2014). It emphasizes the importance of reflexivity and challenges researchers to critically examine their positionality and the impact of their work (Tuck & Yang, 2012; Smith, 1999b).

Land-Based Research Approach

The land-based research approach integrates the land as a living entity and central to the knowledge-creation process, rooted in Indigenous traditions (Simpson, 2014). It emphasizes the importance of place-based learning, where knowledge is derived from direct engagement with the land and its ecosystems (Coulthard, 2014). This approach challenges the disconnection from land inherent in Western methodologies, advocating for research practices that are culturally relevant and environmentally sustainable (Bang et al., 2014). It also recognizes the reciprocal relationship between humans and the land, promoting stewardship and sustainability (Wildcat, 2020; Tuck & Yang, 2014).

Together, Indigenist, relational, decolonial, and land-based research approaches provided a strong framework for exploring Indigenous experiences and perspectives in a manner that is both respectful and transformative. Following these research approaches, we used the following methods in our book. Our methodological approaches recognize that communities are not just subjects of research but active co-researchers to lead us and guide us to transform our research into actions (Datta, 2023).

Indigenous Elder-Led Walk

The Indigenous Elder-Led Walk is a research method that emphasizes experiential learning and intergenerational knowledge transfer, guided by the wisdom of Indigenous elders (Simpson, 2014). During these walks, elders share stories, teachings, and cultural knowledge specific to the land, allowing participants to engage deeply with place-based Indigenous ways of knowing (Archibald, 2008).

This method fosters relationality and respect for the land, reinforcing the connection between community members and their environment (Styres, 2011). It is a decolonizing practice that centers Indigenous voices and challenges Western research paradigms (Johnson et al., 2022; Kimmerer, 2013).

Decolonial individual and collective decolonial conversations are dialogic processes aimed at dismantling colonial power dynamics and fostering alternative, pluriversal ways of knowing (Mignolo & Walsh, 2018). These conversations involve critical reflections on identity, history, and power, emphasizing the need for both personal and communal decolonization (Tuck & Yang, 2012). In collective settings, they serve as a means of reclaiming and revitalizing marginalized knowledge systems while fostering solidarity and mutual understanding (Bhambra, 2014). The process is rooted in the principles of relational accountability and ethical engagement, promoting transformative justice and healing (Simpson, 2017; Smith, 2021).

Researchers' critical reflective learnings are essential for ethical and rigorous research, requiring scholars to examine their positionality, biases, and the power dynamics inherent in the research process (Kovach, 2010b). This practice involves ongoing self-assessment, where researchers interrogate their motivations, the impact of their work on participants, and the broader sociopolitical implications (Smith, 2021). Critical reflection helps researchers to acknowledge and mitigate the influence of their subjectivities, promoting more reflexive and accountable research practices (Wilson, 2008). It also fosters an ethical commitment to decolonize methodologies and ensure the research is community-centered (Chilisa, 2020; Peltier, 2018).

Individual and collective story-sharing is a critical method in research, particularly within Indigenous and decolonial frameworks. It allows for the transmission of cultural knowledge, values, and experiences, fostering a deeper understanding of community histories and identities (Archibald, 2008). This approach emphasizes the relational aspect of knowledge, where stories are co-constructed and shared in a way that honors both individual and collective voices (Kovach, 2010b). Story-sharing is an act of resistance and reclamation, challenging dominant narratives while reinforcing community bonds and intergenerational connections (Thomas, 2005; Wilson, 2008; Simpson, 2017).

In our collaborative partnerships, community elders and knowledge-keepers guided us in our research design, data collection, and analysis. Our book used the aforementioned methods that prioritized relational accountability, meaningful collaborations, transformation, and the integration of Indigenous knowledge systems. We emphasized ethical considerations such as respect, informed consent, reciprocity, collaborative partnerships, transparency, accountability, and the protection of cultural heritage. These principles were crucial for conducting research that was respectful, equitable, and beneficial to Indigenous and local communities. By adhering to these guidelines, we built trust, fostered meaningful relationships, and contributed to the well-being and resilience of Indigenous communities in Bangladesh.

Structure of the Book

In the chapter summaries we provide a concise overview of each chapter, highlighting the main themes and case studies explored. Each chapter is focused on key aspects of Indigenous community–engaged research in Bangladesh, such as ethical research practices, the role of traditional knowledge, research approaches, and the impact of colonial legacies. We integrate case studies and personal narratives from community members throughout the book to offer a rich, contextual understanding of the research topics. These stories serve to interconnect with the theoretical discussions in real-world experiences, showcasing the lived realities, challenges, and resilience of Indigenous communities in Bangladesh.

Chapter 2 focuses on climate change as a global threat that particularly endangers Indigenous peoples in Bangladesh. This research focuses on the Munda Indigenous community in southeastern Bangladesh, which is highly vulnerable. This study uses traditional storytelling and deep listening methods to understand the Munda Indigenous community's unique perspective on disaster resilience. The findings of this chapter suggest a need for a paradigm shift in disaster resilience, guided by decolonial and relational theoretical frameworks. We can provide innovative solutions and practical community-led actions by reconsidering and reclaiming climate change resilience within Indigenous ways of knowing. This chapter aims to transform climate change resilience, governance, policy, and practices in Bangladesh and beyond, offering a fresh perspective on this urgent global issue.

Chapter 3 explores how land and environment have become essential to Manipuri and how they deal with climate change in Sylhet, Bangladesh. The Manipuri's socioeconomic activities developed based on land. In this chapter, we learn how the land is an inseparable part of the community and its customs. According to their culture, the community must have a garden or a forest. For example, the Manipuri cannot celebrate the New Year or cultural festivals without a garden where medicinal herbs and trees grow. This chapter explains how the Maipuri's culture depends on and contributes to maintaining their local ecology. With this base, the community then applies its Indigenous knowledge in managing the community's health, its agricultural practices, and food production. However, due to climate change, the Manipuri face challenges in continuing their livelihoods and culture, not preserving this land-based culture increases the Manipuri's vulnerability and prevents their traditional Indigenous knowledge and community adaptation. In-depth interviews and storytelling methods were used to collect the data.

Chapter 4 explores the impact of climate change on the Marma Indigenous community in Bandarban, a region in southern Bangladesh that is highly susceptible to natural disasters. The community faces increasing threats from flash floods, droughts, and landslides, all of which are intensified by climate change. Using the August 2023 flood as a case study, the chapter reveals the profound effects on the Marma community, who depend on the region's natural resources for their survival. Through narratives from community elders and women, the chapter delves into the flood's causes, its impacts, and the Marma community's coping and adaptation strategies. It concludes with an analysis of the challenges faced by the Indigenous community and offers recommendations for enhancing future resilience.

Chapter 5 examines how the Marma Indigenous community in the CHT leverages its land-based knowledge and practices to enhance climate resilience in the face of escalating environmental challenges. The community's vulnerability to extreme climate events, such as flash floods and water scarcity, has been intensified by unplanned infrastructural development, land grabbing, and deforestation. Adopting a decolonial perspective, the chapter centers Marma voices, addressing the historical injustices and sociopolitical marginalization they face. It underscores the community's traditional land and forest management practices, the detrimental effects of anthropogenic environmental degradation, and the loss of natural resources that jeopardize their livelihoods and cultural heritage. The chapter concludes by presenting community-led strategies for sustainable land, forest, and water resource management.

Chapter 6 examines how climate change affects Munda Indigenous women in coastal Bangladesh, focusing on their specific vulnerabilities to natural disasters like cyclones, floods, and saline intrusion. It underscores the socioeconomic difficulties and gender inequalities these women face, especially when men migrate for work, leaving them more exposed to environmental risks. The chapter also delves into the community-driven coping and adaptation strategies they employ. By amplifying Indigenous voices, particularly those of women, the chapter seeks to urge policymakers to reevaluate the basic human rights of the Munda people and foster more inclusive climate resilience strategies.

Chapter 7 explores the perspectives of Laitu Kheyang Indigenous youth on addressing community disasters in the CHT, Bangladesh. Utilizing story-writing initiatives, the chapter adopts a youth-led approach to explore the causes of flash floods, assess adaptive measures for community resilience, and propose mitigation strategies for government action. Emphasizing a decolonial perspective, the findings reveal that the most effective solutions emerge from within the community. By challenging dominant narratives and amplifying Indigenous youth voices, the chapter illustrates how historical injustices, land dispossession, and sociopolitical marginalization intensify disaster impacts on the Khyeng communities. The chapter advocates for empowering youth in decision-making processes and underscores the importance of sustainable land management that respects Indigenous land rights.

The final chapter will revisit first the changing climate being experienced in Bangladesh by these Indigenous communities, discuss the theoretical and methodological approach of this book's authors and why it is important, and provide a summation of the risks and challenges of Bangladesh Indigenous peoples. Thereafter important adaptations that are happening, as well as those that could happen to reduce their risk and exposure, will be overviewed.

2 The Meanings of Community-Led Disaster Adaptation Strategies

A Case Study with the Munda Indigenous Community in Bangladesh

While Bangladesh is facing a pressing issue with its high susceptibility to the impacts of human-created climate change within the region, there is hope in community-based approaches. Residents of the southern coastal region are not just vulnerable, but highly so, to periodic environmental disasters caused by climate crises. These disasters, including cyclones, storm surges, floods, and other such events, are escalating in frequency, scope, and intensity, posing a significant threat to the community. In this chapter, the Munda Indigenous community,[1] located in the southwestern area of Bangladesh, in the districts of Satkhira and Khulna, has adopted community-led disaster adaptation measures. Individuals must address and manage various natural calamities, such as cyclones, floods, salinity, intrusion, famines, and heatwaves. Roy (2024a); Siddiq et al. (2018); Datta et al. (2024) assert that the prevailing circumstances underscore individuals' significant challenges in sustaining their subsistence near the Sundarbans forest.

Human-created climate change has a more significant impact on Indigenous communities compared to other communities, emphasizing the need for a community-focused strategy. Applying this methodology is essential for formulating grassroots approaches that actively engage and empower the communities affected by climate-related issues. Embracing a community-based approach in catastrophe research is a promising development as it recognizes the diverse viewpoints, ancestral wisdom, and concepts that Indigenous peoples contribute. This approach enhances the efficacy of strategies to mitigate and adapt to climate change and fosters resilience and sustainability by promoting community-level collaboration and engagement. Utilizing a community-based approach holds significant potential for developing effective and contextually tailored solutions that consider Indigenous populations' unique needs and experiences in the context of climate risks. This claim is valid within a global context with a dearth of predictability. The primary goal of this study is to develop location-specific solutions to climate challenges within communities. These solutions should prioritize resource efficiency, cost-effectiveness, and long-term sustainability. These activities facilitate the exploration of creative solutions by formulating these responses Utilizing the decolonial and relational framework has revealed that enhanced collaboration among stakeholders and facilitators is imperative in ensuring the efficacy of

DOI: 10.4324/9781003536345-2

these endeavors. To address the impacts of climate change, it is essential to harness the expertise and customs of the local community. Traditional Indigenous narratives possess a wealth of knowledge and influence within the community, rendering them potentially valuable foundations for practical and empowering health promotion endeavors (Markham et al., 2020). The frequency, scope, and intensity of environmental disasters, particularly those associated with climate change, have escalated the hazards to which individuals are exposed. Integrating efficient disaster risk reduction (DRR) and community resilience-building strategies into development policies, plans, programs, and projects is now widely recognized as crucial for attaining sustainable development goals (SDGs), especially in regions prone to environmental disasters (Imperiale & Vanclay, 2024).

Through this study, the Munda Indigenous community's story shared with policymakers to encourage them to reconsider the fundamental human rights of the Munda people and to provide them with what they are entitled to. Additionally, this chapter focuses on the speech of Indigenous people, and the Indigenous community is the source of all the recommendations. The purpose of this chapter is to accomplish three goals, which are as follows:

- To acquire a comprehensive understanding of the disaster adaptation mechanisms utilized by the Munda community
- To observe the resilience measures from the perspective of a climate change outcome
- To comprehend the perseverance resulting from both physical and cultural health

This study examines how the Munda Indigenous community copes with the impact of climate change through adaptation mechanisms. The results of this study are a constituent element of a broader inquiry to formulate policies and protocols for the Indigenous and minority communities in Bangladesh. We utilized the knowledge acquired from engaging in discussion with community members, who provided their insights on the concepts of climate change, the challenges associated with climate events, and the adaptation strategies they have implemented (Akinola et al., 2019).

In doing so, this chapter has been structured with an introduction containing the research problem and how to solve it, mentioning the purposes of this paper. Then, the positionality of the researchers is essential because, as researchers, we did this research to learn how to benefit from the study and provide benefits to the community. After that, relevant and compatible theoretical directions have been mentioned. Then, findings are received from the related stakeholders, and an analytical discussion is made.

Theoretical Framework and Methodology

In this chapter, we emphasized decolonial and relational theoretical frameworks, reflecting our long-standing connection with the Munda Indigenous communities in the coastal area of Bangladesh. These frameworks were essential in examining the interplay between climate change resilience and the altered physical and cultural health conditions resulting from climate change. Both frameworks highlight the importance of land, nature, and sustainability within these communities. Our approach helped dismantle and disrupt prevailing narratives and systems that have historically oppressed Indigenous and other vulnerable populations. We recognized that conventional research methods often perpetuate unequal power dynamics by treating Indigenous knowledge and peoples as objects of study, thus exploiting them. The primary aim of our decolonial research paradigm was to prioritize Indigenous voices, knowledge systems, and worldviews while critically analyzing the impact of colonialism on the research process (Wigtil et al., 2022; Mungazi, 1986).

By adhering to decolonial and relational theoretical frameworks, we regarded the land as a vital relationship—a powerful conduit for transmitting cultural heritage, fostering collective identity, and inspiring a commitment to environmental stewardship among younger generations. Land-based learning facilitated the development of deep respect for nature, cultural appreciation, and collective responsibility in addressing the impacts of climate change (Greene, 2021). These frameworks, coupled with concepts of nature, land, and sustainability, were instrumental in understanding climate change resilience, particularly in the context of disaster adaptation strategies.

We employed individual and collective conversations as our primary research method, achieved through deep listening, traditional story-sharing, land-based discourse, and reflective thinking. Deep listening was a crucial approach, essential for demonstrating respect for Indigenous perspectives. Storytelling played a vital role in reconnecting with land-based learning, as Indigenous narratives are intrinsically tied to the land, language, and essence of the people. Research stories were gathered during two distinct periods: August to September 2021 and July to August 2023. These stories, collected from the Munda community elders and knowledge-keepers in Shyamnagar Upazila, Satkhira District, revealed the community as a repository of unique perspectives shaped by traditional wisdom, experiential understanding, and deep environmental connections. According to Tubridy et al. (2022), our findings illuminated the complex interrelationships between communities and their environments, highlighting how recognizing Indigenous perspectives can enhance the effectiveness of climate-related policies and strategies, aligning them more closely with the needs and aspirations of those most affected by climate change.

Finally, we used thematic analysis to evaluate the outcomes of our research. This method allowed us to focus more effectively and enhance the quality of our academic inquiry (Akinola et al., 2019; Mathbor, 2007). Our efforts were directed at fostering self-determination within communities by promoting

Indigenist and land-based methodologies that strengthen resilience (Strong, 2005; Topna, 2002). Throughout the research, we maintained a deep respect and gratitude for the community elders and knowledge guardians. The privacy of the Munda Indigenous community, a marginalized group, was preserved, ensuring that their rights and dignity were upheld.

Findings: Learning Reflections from the Munda Community of Bangladesh

This section addresses the impact of climate change, the community's adaptation strategies to climate change, and proposed solutions. Following periods of deep listening during the sharing of stories and subsequent review of the recordings, we reflected on the knowledge we acquired in our findings. Reflection facilitated our acquisition of knowledge in a responsible manner. Listening to recorded interviews provides a unique perspective compared to reading written transcripts. The act of listening engages the listener's emotions in a heightened manner. Indigenous women shared their opinion and experiences in our educational narratives, which sometimes proved challenging and emotionally impactful. When we selected quotes and personal stories, personal preferences played a key role. We focused on quotes that left a lasting impression on us, were easily remembered, and held significance for our analysis and memorization. We continuously listened to and learned from the Munda community to develop main themes and subthemes (Table 2.1).

Impact of Climate Change on the Munda Community

Climate change has significantly impacted the Munda community, resulting in unpredictable and frequent human-created environmental disasters and their corresponding effects. Prominent ramifications encompass:

Table 2.1 Main themes and subthemes of the chapter

1 Impact of Climate Change		*2 Adaptation Strategy*	*3 Proposed Solutions*
Livelihood • Drinking Water Crisis • Heatwave Education • Decreasing Trend in Mobility • Improper health facilities • Land Rights and leaving Habitat • Substance Abuse	Environmental • Salinity • Floods • Cyclone	• Shelter • Saline Resilient Production • Loan Taking • Water Management	• Land Rights and housing. • Access to Education • Health Complex Facilities • Financial Support • Employability

Crisis of Drinking Water

Collecting fresh water is difficult due to the arid summer season and high saline levels. A female participant expressed an issue with the quality of drinking water:

> By employing filtration, we obtain potable water from the freshwater reservoir. However, we must travel considerable distances to procure water in floods. Water is a necessity and entitlement; nonetheless, acquiring freshwater from coastal regions poses significant challenges. We experienced water scarcity for half of the year.

Another male respondent from the same locality echoed the same voice. During the speech, the speaker specifically mentioned the increasing salinity levels in the water, causing contamination of shrimp farm pond water. The Sundari tree, a famous tree, has also been dying.

Heatwave

Like the other residents, the Munda Indigenous community has experienced heatwaves. Nevertheless, this Indigenous population is particularly susceptible because of their long-standing rural ways of existence, which are intricately intertwined with the natural environment. Another female participant stated, "July and August typically experience the highest temperatures throughout the year, but there is a significant annual increase." She attributed the cause of heatwaves to environmental changes and elevated carbon dioxide levels. Heatwaves also cause salinity. Multiple factors contribute to the problem, such as overpopulation, limited access to the Sundarbans forest for wood collection, which affects other livelihood activities, inability to fish due to salinity, scarcity of fish, and significant levels of shrimp aquaculture. One respondent mentioned, "The weather becomes hot due to '*gher*.' Where there is '*gher*, it is hot'; on the contrary, where there is paddy cultivation, the weather is average."

Reduced Mobility

Climate change also impacts people's income in this area, raising the unemployment rate. Unemployment is associated with working in remote places and relying on borrowing funds. The Munda Indigenous communities typically resides near the river, engaging in the capture of fish and crabs, as well as in the Sundarbans forest, where they gather honey and herbs. The Sundarbans and its adjacent rivers are inaccessible for six months, leaving individuals with no activities to engage in. A male participant stated that he frequently visits the brickfields, but if he finds his way to the forest and the police catch him, they will set fire to the net and boat. Given the harsh conditions of the environment, there are no viable alternatives for employment. Consequently, males and adolescents seek different means of generating income in the external environment.

They engage in labor at the brick field. Another female participant said the Indigenous people borrowed funds from others during calamities and unemployment. Consequently, they must once again seek a loan to reimburse the funds. This situation perpetuates the cycle of borrowing.

Education

Education is a vital component for the survival of Indigenous communities. However, within the Munda living areas, the pursuit of education presents more significant difficulties, considerable distances, and inconveniences for the local population. In a calamity, the children cease attending school due to the submergence of roadways. One participant stated the Indigenous children could not attend school during the monsoon season due to the damaged road. Conversely, if a shelter necessitates their school to transform into a shelter center in a crisis, it impedes their education. Furthermore, the school's location makes it difficult for young children to commute alone. Consequently, older individuals are necessary to accompany the children, which poses a significant difficulty for their parents. Girls are facing increasing challenges in accessing education. They are compelled to enter marriage at a young age and abandon their education. A female student from a Munda school conveyed her viewpoint that her parents expeditiously arranged her marriage with a girl's successful completion of the secondary school certificate examinations. Hence, the Munda Indigenous community faces challenges in accessing education, which is further exacerbated by the impacts of climate change.

Improper Health Facilities

The Munda people are experiencing extreme poverty and are unable to afford three meals each day, which makes them susceptible to illness. Another participant said, "Our food supply is inadequate, resulting in malnutrition." Indigenous populations residing along the coast experience poverty and lack access to healthcare facilities in their regions. During an emergency, individuals must travel to the town for medical care, which incurs high costs and lacks convenience. This problem is exacerbated during the flood when individuals cannot navigate their surroundings. In the aftermath of each calamity, what is the survival duration based solely on dry food and water consumption? "During that period, individuals experience anxiety and illness," stated a participant. According to the same respondent, individuals perish due to inadequate medical care. Amid the catastrophe, they experienced complete loss, and the recollection of this event caused them to be deeply distressed.

Local, National Change, and Colonialism

Land Rights and Habitat Change

The Munda Indigenous population migrated to this region three centuries ago. For years, they have resided in Munda Indigenous villages. Nevertheless,

Indigenous women have expressed their anger over the lack of land rights they possess due to the imposition of colonial ideology. One participant described a sense of distress due to the absence of land ownership. "Individuals have a modest parcel of land in every location. Nevertheless, we lack ownership of our land; we reside on land belonging to others. Individuals are required to reside in the residences of others. We are about still determining our destination. We need more financial resources to purchase our land". Furthermore, irrespective of the size of the family, they must reside in a solitary room. Their residence needs adequate sanitation and bathing facilities. One participant highlighted the several challenges arising from climate change. The toilet and sanitary facilities are unsatisfactory and could be improved. The residence they reside in could not be more conducive to disasters. During the arid season, their dwelling deteriorates due to its mud composition, and they lack the financial means to reconstruct it. Another female participant commented that constructing a mud house with saline water was unsuitable. During the dry season, it undergoes decomposition. According to another woman respondent, they evacuated their residences during the catastrophe.

Substance Abuse

Recently, there has been a trend among men and boys to labor in brickfields due to the lack of employment opportunities in their village. Consequently, individuals of all age groups are engaging in smoking, drinking, and developing addictions. Due to their remote location, the adults cannot take any action against them. Additionally, the children acquire this knowledge by observing and imitating others. One respondent stated that they carry both local and Indian booze. Moreover, they allocate all their earnings toward purchasing alcoholic beverages. According to the same respondent, upon their return home, they engage in physical violence against their children and spouse. The Munda community exhibits a customary practice of consuming alcoholic beverages throughout their festivals and festivities. Nevertheless, it has now become ordinary, resulting in psychological distress for their family due to their frustration.

Our analysis revealed that the impact on livelihood is especially significant in coastal regions and is a primary driver for climate-induced relocation. Bangladesh serves as a compelling illustration. The coastal regions' susceptibility exacerbates the socioeconomic consequences of climate change, directly impacting the ability of communities to maintain their livelihoods. The community has articulated that the escalation of sea levels; occurrences of severe weather phenomena such as excessive rainfall, drought, cyclones, tornadoes, and other similar events; and alterations in environmental patterns all contribute to the degradation of livelihoods. These factors impede conventional economic activities, including agriculture and fishing. As a result, communities are forced to migrate or relocate in search of sustainable alternatives.

Understanding the intricate links between climate change and the loss of livelihoods is crucial for developing effective ways to lessen its adverse effects and protect the well-being and ability to adapt to communities living in vulnerable areas.

Adaptation Strategies

In the Munda Indigenous community villages, Indigenous people have developed their own coping and adaptation strategies, some of which benefit the community and some of which make them more vulnerable. Here, we will discuss their adaptation strategies.

Shelter

In Bangladesh, the enduring consequences of colonization persistently exert a substantial influence on Indigenous groups, with a particular emphasis on the Munda Indigenous groups. Past colonization's enduring effects have significantly influenced communities' sociocultural, economic, and political features. The impact of this influence is seen in the gradual decline of customary traditions, the forced relocation of Indigenous communities, and the marginalization of their perspectives within broader social frameworks. In the event of a disaster, individuals must promptly seek refuge in a nearby shelter due to the absence of a secure living environment. The Munda Indigenous group, however, does not seek refuge in shelters. One participant expressed their inability to visit the government's cyclone center due to its considerable distance from their current location. Additionally, they observed that the facility had become occupied well before her arrival. Hence, they opt to remain at home despite the inadequate structural integrity of their dwellings to ensure their safety. Hence, their approach entails remaining within the confines of their residence. If the situation worsens, they seek refuge within the school premises. According to another participant, significant documents were taken, and all personal goods were left within the residence. According to the same respondent, there were instances where individuals sought refuge on the road due to the refusal of the local chairman of the union to grant permission for them to seek shelter within the school. The occurrence can be attributed to their disparate socioeconomic standing. Consequently, it is imperative to acknowledge and tackle the consequences of persistent colonialism to promote sustainable development and safeguard the abundant cultural legacy of these underprivileged people in Bangladesh.

Saline-Prone Crop

By embracing saline-resistant paddy cultivation, i.e., IRRI Hybrid, typen23, 51 paddy, a new livelihood pattern is emerging, transforming people's lives. They also produce saline-resilient fruit, such as sapodilla plum, elephant wood apple, and coconut. Through this approach, individuals can adjust to evolving circumstances. They also try to produce other crops to adapt to the changed environment. They have also become habituated to drinking a small quantity of saline water. On the contrary, one respondent mentioned that they have not received guest birds in winter for the past three decades.

Taking a Loan

Individuals frequently experience complete loss after a calamity. They need sustenance, and their dwellings need to be more disaster-protected. At this point, people require assistance but do not receive it or any form of relief. The government assists with other settlements but does not need help to reach them due to their geographical position. One participant stated that they visited the adjacent regions, but access was needed to access the most remote locations where assistance was offered. The occurrence was a result of their diminished significance. The government has numerous amenities, but all receive preferential treatment from the chairman and members. Consequently, we must shoulder the burden of people and nongovernmental organizations. When the local hospital cannot offer treatment, individuals also get loans to support their family members afflicted with severe illnesses. The circumstances have evolved so individuals can sustain themselves only by obtaining loans.

Water Management

The primary challenge faced by the Munda Indigenous community pertains to the availability of water resources. The region has one or two freshwater ponds serving water sources for various purposes, such as cooking, drinking, and daily activities. Certain settlements are equipped with filters to separate the pond water. Nevertheless, a predicament develops when the deionized water infiltrates the freshwater reservoir. The villagers use their utmost efforts to preserve and venerate the freshwater pond as a divine entity. They have explicitly stated that they do not pollute the pond that holds fresh water and emphasized that their lives will be irreparably damaged if the ponds get polluted. The individuals independently construct a water collection and conservation system in collaboration with organizations such as Sundarban Adivashi Munda Sangstha (SAMS) and the United Nations Development Programme (UNDP). Another Munda participant stated, "We collect rainwater from the UNDP-provided tank and store it. One year ago, we acquired a tank." In the event of a crisis, the task of water collection becomes challenging, as evidenced by reports stating that in the event of a flood, individuals are compelled to travel considerable distances by boat to procure potable water. During the monsoon season, rainwater is stored in a reservoir. Nevertheless, some of them reside in extreme destitution to purchase this tank. It can be hypothesized that the Munda community is facing a severe water problem, necessitating the implementation of remedies to facilitate their adaptability.

Suggested Solutions

Implementing certain modifications within the societal framework is imperative to improving the circumstances of coastal Indigenous populations. This chapter incorporates a selection of solutions proposed by Munda community members. These recommendations originated from their period of sharing stories.

Land Rights and Housing

The government must provide land to Indigenous communities nationwide. "If we possessed our land, it would be more advantageous," expressed each participant. In addition, they desire to possess property and residences. Additionally, they require aid in constructing their dwelling, which will ensure their safety in the event of floods or storms. Several participants also believed that relocating to a native region would enable them to sustain themselves through agricultural activities and farming. The mention of land-based solutions is not unexpected. One respondent wanted to accumulate funds and purchase a small plot of land to construct a house. The local government does not embrace Indigenous individuals in its forum, and it is imperative to include Indigenous individuals as community leaders to empower marginalized communities. Several participants believed that a cyclone center in their vicinity would be advantageous. The Munda Indigenous community would have significantly benefited from constructing cyclone centers near their village locations.

Access to Education

The Munda Indigenous group exhibits a strong passion for education, although it encounters several challenges. One participant expressed that the primary school is far from their residence. The need for schools and universities nearby to provide for their children is considered a fundamental human right. Due to the inadequate construction of specific locations, respondents proposed that repairing the roads would facilitate their children's access to school. Additionally, they stated that they do not receive any financial assistance for their children's education. If they receive financial aid for schooling, their children are required to work only if they attend school. The government must ensure the stipend is also allocated to the community's children.

Health Complex Facilities

In impoverished coastal regions, individuals cannot access medical care, and the healthcare facility is far from their homes. Here, individuals perish due to inadequate medical care. "We require a physician and a medical facility," stated a participant. Additionally, the provision of free or low-cost healthcare necessitates the establishment of government-operated hospitals.

Financial Support

The Munda village narrative demonstrates that even a modest sum of money can significantly impact. A participant expressed that receiving a monthly sum of 300tk would greatly benefit them. The Disaster Vulnerable Group Feeding Program (VGF), a governmental financing initiative within the locality where they lack access, is vital. The authorities should prioritize the inclusion of local authorities in a governmental facility-providing system.

Employability

Indigenous people and mainstream communities in the coastal area of Satkhira need more working facilities. "It would be advantageous if the government engaged us in a revenue-generating endeavor," stated a participant. He also said their involvement in financially lucrative work would benefit us. The government also provides Indigenous individuals with additional employment quotas, albeit with significant limitations. Another female participant said that creating a connection is necessary for finding a job, and the government can help. It also suggested that to stop "gher" farming, agriculture may generate employability.

The proposed remedies potentially impact the livelihoods of the Indigenous coastal communities significantly. This would be exceedingly challenging, as minority individuals are perceived as a subordinate to the majority population. Furthermore, colonization can influence individuals' cognitive processes and lead them to believe in repressing the rights of minority groups. Authorities from various domains must work together to bring about these reforms.

Discussion and Conclusion

The locality of the Munda community is prone to human-created environmental disasters such as cyclones, floods, and erosion due to factors like sea-level rise, droughts, and saline intrusions. Climate change has numerous adverse effects on humans, animals, and ecosystems, resulting in poverty, severe scarcity of food and water, unemployment, saltwater incursions, drug addiction, and physical and mental health problems. This chapter examines the ramifications of climate change, the tactics employed for adaptation, and the community's proposed remedies from the Munda community. This report identifies three main findings: the role of Indigenous leaders, restructuring adaptation measures, and rethinking climate policy.

Recognizing Decision-making Authority

The study demonstrates the systematic exclusion of Indigenous individuals from all rights and amenities. The Indigenous storytellers expressed a similar sentiment. According to Bronen (2011), individuals are prohibited from engaging in local decision-making and service provision in a manner that confers advantages to them. Bronen (2011) contends that adopting a system-thinking approach to adaptation would facilitate a shift from an economic-centric framework to a people-centric framework that prioritizes human rights, local involvement, input, and decision-making. This approach would also highlight the social, political, and economic disparities our existing system generates (Alverez et al., 2020). Hence, we propose that including the Munda Indigenous population in leadership positions inside Bangladesh is imperative to address their issues and challenges. Indigenous leaders from the Munda community have a direct experiential understanding of their distinct cultural customs and sustainable methodologies, facilitating the customization of policies to align with their specific requirements. The inclusion of Munda Indigenous individuals in leadership positions plays a crucial role in the advancement of cultural preservation, the cultivation of community resilience, and the attainment of equitable development.

Recognizing Adaptation Strategy and Policy

The research showcases the existing adaptive mechanisms employed by the Munda Indigenous population. Individuals undertake their tasks independently in their domains, as they do not get support or aid from external entities. They arrange for a catastrophe and beseech them to remain at their residence regardless of the outcome. Due to their social rank and the hassle associated with shelters, they cannot access any refuge. They get loans, exacerbating their predicament. According to Rabbani et al. (2018), there is concern over the vulnerability of sustainable disaster response and recovery plans, coastal management, resident relocation, and livelihood loss. Hence, it is imperative to restructure adaptation measures for the Munda Indigenous people in Bangladesh to tackle their unique vulnerabilities and cultural issues effectively. By prioritizing these methods tailored to the unique requirements of the Munda community, we can enhance our ability to withstand and recover from the effects of climate change. By reorganizing adaptation initiatives, it is possible to advance sustainable practices based on Munda traditions, cultivating a resilient and culturally attuned approach to addressing environmental concerns. The study has observed insufficient existing measures to address climate change. Indigenous communities are affected by the mismanagement, therefore there is a need for more awareness. There is a need to alter this thinking and policy. Furthermore, it is imperative to undertake endeavors to achieve comprehensive disaster management, flood control, enhanced resilience of vulnerable groups in both rural and urban areas, migration, and other significant concerns. To tackle these problems for the disadvantaged, it is necessary to implement innovative, open, and easily understandable solutions (Shammin et al., 2022). Hence, it is imperative to reassess climate change adaptation measures to effectively address the distinct vulnerabilities faced by Munda Indigenous communities in Bangladesh. This examination should consider the interplay between cultural practices and environmental transformations

(Maldonado et al., 2014). Integrating the land-based perspectives of the Munda Indigenous group is crucial for improving the alignment between climate change policy and their traditional knowledge and resilience. The requirement to defend the rights and well-being of Munda Indigenous populations experiencing the growing consequences of climate change in Bangladesh necessitates the inclusion of inclusivity in adaptation methods (McVittie et al., 2019). Salinity and scarcity of fresh water are significant consequences of climate change in certain areas. Many illnesses stem from that. Planning for daily water consumption is not advisable due to the high costs involved. Consequently, their reliance is contingent upon the existing water supply mechanism. The study highlights the high susceptibility of the coastal population, leading to devastating casualties and property damage (Asugeni et al., 2017). They suffer from a range of dermatological conditions resulting from factors such as salinity and malnutrition, and they lack access to medical professionals in their locality. In addition, they are unable to conduct fundamental health examinations. In an emergency, they seek treatment from a neighboring quack, which is neither reliable nor authorized.

Recognizing Munda Indigenous Community's Land Right

The Munda Indigenous community lacks ownership or legal entitlement to their property. Their residences are unsuitable and must be more suitable for habitation, necessitating extended periods of absence from home to seek employment. Many of the Indigenous community members do not have access to employment, resulting in a lack of essential resources such as drinking water, food, and employment. Numerous individuals expressed a lack of a stable future and alleged that the administration disregarded their fundamental human rights. Their lives are perplexing. Their lives have been left stuck due to poverty, unemployment, water problems, salinity, and different human-created environmental disasters. Furthermore, inequities and unequal access to government benefits during calamities have significantly impacted their quality of life. In addition, inequalities in various amenities, exclusion, and societal inequity have caused profound distress in their existence (Brown et al., 2016).

The Munda Indigenous community has been adversely affected by both continuous colonization and climate change. It is crucial to incorporate the perspectives of the Munda Indigenous population to ensure that climate change adaptation plans in Bangladesh are specifically tailored to address their unique experiences and challenges. Embracing an inclusive approach is essential to effectively address the vulnerabilities of the Munda community and strengthen their resilience. This approach can also facilitate the implementation of climate change initiatives that are effective, culturally sensitive, and respectful of the customs of the Munda Indigenous people while promoting sustainable solutions. Bangladesh can effectively address the challenges posed by climate change by leveraging the land-based knowledge and traditions of the Munda population. We can develop effective, practical strategies to address climate-related issues by understanding the complex relationship between the community and the environment. Bangladesh stands to benefit significantly from the expertise of the Munda community and can work together toward a future characterized by improved opportunities and resilience.

Note

1 The Munda Indigenous communities are a distinct ethnic group that lives within the Sundarbans mangrove forest reserve. They came from India and migrated to Bangladesh 600 years ago, where they have since established a permanent presence. The Munda people are in the Khulna, Jessore, and Satkhira Zila, close to the Sundarbans and the Joypurhat areas. The inhabitants of the Sundarbans mangrove forest rely heavily on the ecological community for their sustenance and livelihood. According to a resident scholar,

Munda in Bangladesh are considered tribes. They are also locally known as Bruno, Kooli, Saotal, and Sardar. No one can say exactly when and how they came to former East Bengal, today's Bangladesh, and how they were identified. They came from the Ranchi of Jharkhand and Bankura, Nagpur, Birbhum, Purulia, Santal Pargana, and Mednipur districts of India's West Bengal. First, the king of Naldanga in Kaliganj of Jhenidah (the then Jessore) district brought the Munda people here. They worked as clubmen (locally known as Lathial) in the house of kings; secondly, after the failure of the Shantal rebellion in 1850, British Ruller brought them here to cultivate the plants of Indigo. Thirdly, the landlord of the South Western region brought them to Bangladesh to cut the Sundarban and embark on a plan to prevent saline water from entering the cultivable land and to make human habitations in the places near the Sundarbans.

3 Indigenous-Led Climate Change Adaptation from and within the Manipuri Indigenous Community-Based Perspective

This chapter explores the impacts of climate change, concurrently with development and land-use change, on the Manipuri, an Indigenous people in Bangladesh. The Manipuri is one of the Indigenous groups living in the Sylhet division in the northeastern region of Bangladesh. The Manipuri speak the Manipuri language, also known as Meetei-lon. Manipuri is one of India's national languages. Moreover, the Manipuri language or Meetei-lon is the official language of Manipur and Assam in India. The language has a script (Meetei-mayek) that is nearly 3,000 years old. The Manipuri language belongs to the Tibetan-Burman family. The Manipuri call themselves Meetei. Meetei became Manipuri in the 18th century after the adoption of Hinduism. Moreover, Kangleipak became Manipur in the same period. So, Manipuri and Manipur terms are recent according to linguistic, historical, cultural, and archaeological considerations. The Manipuri people follow the ancient Sanamahi religion, as well as Hinduism. It is not certain how many Manipuri live in Bangladesh. According to the population census of the Bangladesh government in 1991, the Manipuri population is 24,882 (Singha, 2022). However, another study reveals that Bangladesh's Manipuri population is 7,372 (Thongam, 2008, p. 18). The average literacy rate is 25% (Thongam, 2008, pp. 15–18).

Indigenous land is integral to their identity, culture, and daily livelihoods. Manipuri follow Indigenous knowledge of nature, such as the earth, water, and the sun. Moreover, Manipuri culture and religious festivals are related to nature. Nature and the environment are integral parts of Manipuri's daily livelihoods. The culture and beliefs of Indigenous people are related to the land (Akter et al., 2019; Newing, 2010). A recent study claims that Indigenous elders and knowledge-keepers' (IEKK) land-based statements play a crucial role in environmental sustainability, such as physical, spiritual, emotional, and cultural improvement; this is in part due to the fact Indigenous people's lives are based on land (Datta, 2023). The culture and language of Indigenous communities are closely centered on the land. Land-based communities such as Canadian First Nations, Metis, and Inuit background people have diverse local vocabulary based on land that highlights their cultural relationship with land.

There is a live connection between customary land and history, language, knowledge, and health (Redvers, 2020a). Similarly in the Manipuri community,

DOI: 10.4324/9781003536345-3

their livelihoods are still closely connected to nature and the environment. However, the Manipuri face land-based challenges, mostly manmade, and some of them are created naturally in Bangladesh. According to Ashrafuzzaman (2014), "Indigenous values, traditions, and livelihood were neglected in the development policies because the policies aimed solely at economic development, not cultural development. Rather, many development policies and programs cut their rights to their ancestral land" (p. 10). The author also suggests, "The human rights of Indigenous people are violated continuously all over the world because of land conflicts, which result in threats and deaths of individuals, disappearances of whole communities, and forced changes in cultural habits (forced assimilation) and socio-economic relations" (pp. 9–10). Ashrafuzzaman also suggests that Indigenous communities of Bangladesh face land-related issues. The government and its policymaking have not focused on local cultural significance and occupation-related issues. Further, Faisal, Singha Ahmmed, and Singh (2007) conducted a study of the socioeconomic condition of the Manipuri. The study mentions changes in food habits due to interaction with the mainstream community and modernization. However, this study, as well as that of Ashrafuzzaman, did not focus on climate change and its impact on Manipuri and their traditional food production. Singha (2015) argues about the importance of Manipuri housing design and how it relates to the environment and belief system. However, the study highlighted less on how climate change impacts the local Manipuri housing design.

Indigenous communities are victims of ecosystem climate variation. Nowadays, they are vulnerable to climate change because they are not well prepared to face the challenge. Climate change impacts traditional Manipuri livelihoods, culture, and food habits. While there are some studies on how climate is changing in the region (Akter et al. 2019; Islam et al. 2006), there have only been a few studying the impact on the Sylhet region, and even fewer on Indigenous people (Mustajib, 2020). There has been little literature on the perspective of climate change and the challenges faced by the Manipuri in Bangladesh. Community-based adaptation (CBA) plays an important role in minimizing the impacts of local climate change (Kirkby et al. 2015). Collaborative community engaged methods are essential in responding to land issues (Hunsberger et al. 2017).

There has been less research on Manipuri and climate change adaptation issues, though climate change is an integral part of the community. Manipuri identity and culture are based on land, an integral part of their daily livelihoods. However, their land is shrinking quickly due to socioeconomic challenges mainly brought on by manmade and human-created environmental disasters. This leads to identity and health impacts for the community. We conducted fieldwork with an in-depth interview method with resource persons and local Manipuri communities in the Sylhet region in 2023 to explore the connection between the Manipuri and land-based culture, Indigenous knowledge, and environmental challenges.

This research is unique because it explores how the Manipuri contribute to preserving the environment while trying to adapt to climate change, focusing on CBA knowledge, particularly Indigenous knowledge. First, we describe my positionality and research methodology. Then we recount our findings starting with the centrality of earth to the Manipuri and their culture. Then, the paper examines how climate change impacts the livelihoods of the Manipuri. For example, the paper investigates how climate change affects the local environment, ancestral land, Indigenous food substances, and Indigenous community knowledge. Then, the paper explores the challenges experienced by the Manipuri, such as the loss of land, deforestation, and resulting socioeconomic impacts worsened by climate change. This study in this chapter attempts to create a synergy between Manipuri and relevant stakeholders to minimize the effects of future disasters, focusing on land-based community viewpoints. Finally, this paper contributes new scholarly knowledge to the current literature. The chapter is divided into five sections: researcher positionality, theoretical framework and methodology, findings, discussion, and conclusion.

Researcher Positionality

Following the relational theoretical framework, this research was led by an Indigenous scholar, Dr. Singha (fourth author), a Manipuri Indigenous person. Dr. Singha holds a deep sense of responsibility toward the community, particularly Indigenous communities, and is committed to protecting nature and the environment, as entrusted by *Ima* (earth or mother). When presented with the opportunity to conduct this research, Dr. Singha readily agreed, recognizing its importance for the Manipuri community and future generations.

Dr. Singha previously wrote a graduate thesis on the beliefs and rituals of the Manipuri, during which he realized the intrinsic connection between Manipuri culture and the earth and nature. This understanding deepened during the writing of a master's thesis focused on the housing knowledge of Manipuri, sparking an interest in how Indigenous housing design relates to land, nature, and the environment. Recently, climate change has significantly impacted the Manipuri community, similar to other Indigenous communities in Bangladesh. This research has enriched Dr. Singha's understanding of the community-based climate adaptation practices that Manipuri have cultivated over generations.

Climate change affects the socioeconomic and environmental sectors of most communities in Bangladesh, particularly Indigenous communities like Manipuri, whose culture and livelihoods are increasingly threatened. The findings of this research can be disseminated to the broader community, raising awareness about climate change and how it can be adapted while respecting local culture, Indigenous knowledge, and the environment.

Dr. Singha aspires to become a leading voice within the Manipuri community, informing stakeholders and collaborating with the community in designing and decision-making processes. This work is vital for the future of the Manipuri people and the wider Bangladeshi community. Dr. Singha hopes that this research will contribute to a deeper understanding of the

Manipuri community and his connection to the environment, both at the local and national levels.

Theoretical Framework and Methodology

Following a relational theoretical framework, we used in-depth storytelling methods to collect knowledge from co-researchers, local elders, and community members. Co-researchers (fellow authors of this book) have become a part of the research team because their viewpoints are emphasized (Datta, 2024a). Researching tribal stories requires consideration of Indigenous peoples' experienced ecosystem, self-determination, hopes, and aspirations from a local worldview (Kovach, 2021, p. 97). We applied storytelling methods to gather our data. We selected respondents who have better knowledge than ordinary people. Our respondents are regarded by us as co-researchers. We shared our stories and experiences about climate change, its history, impacts, and adaptation. We applied storytelling techniques aligned with local culture to gather statements from the community. The following arguments claim the significance of key respondents. For example, a key respondent can easily express their worldviews. We asked a few questions but allowed the interviewee "to respond freely, with the interviewer simply responding to points that seem worthy of being followed up" (Bryman, 2016, p. 470).

We consider it a storytelling method because we tried to learn from the community and their experiences. "An in-depth interview aims to obtain as full and unbiased an account as possible of the participant's perspective on the research topic, and the researcher's task is to use every means at their disposal to aid this" (Lewis, 2003, p. 158).

We conducted this fieldwork in several stages. First, in August 2023 we interviewed among the Nayabazar Manipuri para in Sylhet City, Bangladesh. They are urban Manipuri, but socioeconomically, they are disadvantaged. There are 40 families in the para with a total of 200 people. Then, we conducted another field investigation in October and December 2023 in the Kamalganj, Moulvibazar District where most Manipuri people live. This fieldwork enriched our earlier knowledge about Manipuri and their climate change–related knowledge. We revisited the research area in Kamalganj in April 2024 and updated the findings. More recently, in 2024, we did another field investigation, followed by a remote interview. We collected the data over the phone and recorded the respondent's statements.

We went to the Manipuri para in Sylhet with prior permission and followed the cultural manners and norms. Moreover, Dr. Singha initially connected with the key respondents during his PhD fieldwork between 2016 and 2017. We obtained consent from respondents before starting every interview. We also avoided culturally sensitive questions during the interview. We shared our draft article with respondents for their comments and suggestions. Later, we incorporated their valuable comments into the paper. Our overarching research questions included the following:

How climate change affects Manipuri livelihoods;
How climate change impacts Indigenous knowledge and food production in the community;
How land, nature, and environment are related to the Manipuri culture and identity;
How colonization creates landless people and an identity crisis among the communities

Findings: The Centrality of the Earth for the Manipuri

During the second field trip, one of the respondents mentioned that the earth is functioning similarly to his mother. We live on the earth (mother) and are responsible for looking after it. Nilachand stated: *Mai, ising amadi nuksit (fire, water and air). Asi ahum ase yaondana hingba yade malem thawai panba pumnamak. Asomdana, Leibak ase Imani. Akhoi imada tanjaraga hingbani malem asida* (Manipuri). [Living beings cannot survive without fire, water, and air. Then, earth is our mother. We live on the earth, so we must take care of Mother]. The Manipuri community survives with its land-based culture. The community has a strong respect for the earth. The Manipuri believe they live on the earth, so it is their responsibility to look after it. It also indicates that caring for fire, water, and air is a part of nature. The first key respondent provided the following statement about climate change. Nilachand explained, *Prokriti bhalo na thakle, amra bhalo thakte parbona Amader nijeder jannya prokritike bhalo rakhte hobe* (Bengali). [If nature is in good shape, we can live peacefully. We are part of nature. So, we must maintain a natural intake for our benefit]. Nilachand mentions that one should not disturb nature. Nature has its system of management. Human beings are part of nature. Being a part of nature, one should not destroy nature. Humans should coexist with nature. The statement indicates the Manipuri's spiritual connection to nature. However, the community observes that climate change affects their livelihoods in the following ways.

Climate and Development Impacts on the Manipuri

Impact of Modern Knowledge and Technology on Indigenous Food

The Manipuri are facing the extinction of their traditional food. From the Manipuri's experience *yongchak* (Parkia Timoriana) trees are dying slowly in the area. Historically there were plenty of yongchak trees in the area. The tree produces beans, which are called "yongchak" in Manipuri. Yongchak is one of Manipuri's favorite food substances. Yongchak is also a medicinal fruit for the community. But now, Yongchak has become a rare tree because most of them have died for unknown reasons. Several respondents recounted there are many mobile companies in Bangladesh. They are mobile towers across the country. The community suspects that mobile towers release an electromagnetic wave that kills yongchak trees. In the end, the community has lost an essential Indigenous food substance from nature. Recently, hybrid seeds have become available in the market but are unsuitable for the local land and environment. For example, hybrid varieties produce better yields but are less resistant to climate change, resulting in lost crops due to drought and flood. Nilachand stated,

> *Age amader bab dadara lota pota diya ruger chikitsa korto, akhon segulo khuje pawa muskil hoye gese. 20 theke 25 ta herbal gas, lota hariye gese bolbo ami, karon segulo akhon sochorachor dekhina. Jolobayu paribartan akta karon hisabe dekhi* (Bengali). [Herbs play an essential role in Manipuri culture and livelihoods. The herbs are part of the community's daily life, and we were able to get those herbal items in the area quickly. However, the respondent claims that between 20 and 25 herbal and medicinal plants have disappeared from the Kamalganj area].

Community members explained that something needs to be fixed with climate change, or the use of modern chemicals in agriculture, locally and in other places.

> *Piajer dam akhon beshi, amra sadharonto khaina. Amader akta herb ase ataka yenam bole thaki amra. Piyajer poribarte amra, yenam babohar kore thaki* (Bengali). [Manipuri usually uses one kind of herb, which is known as yam (chives) instead of onion. Recently, onion prices have increased significantly, making buying difficult. Sometimes, ordinary people cannot buy it.]

Most Manipuri do not buy onions because they use *yenam*. Nowadays, the price of one kilogram of onion is US$1.5, which is beyond the buying capacity of the ordinary Manipuri people.

Another critical point is that the Manipuri use *yenam* leaves during cooking. The Manipuri cannot imagine being without *yenam* because it is an essential cooking ingredient. The community plants *yenam* in the backyard or rooftop. The community believes that it contains antioxidants. The Manipuri also eat mushrooms as a part of their vegetables. They eat two types of mushrooms: *charuyen* and *uyen*. *Charuyen* usually grows on dead straw and hay, and *uyen* on dead trees. They do not eat other mushrooms. Other edible mushrooms may exist, but the community does not eat those unknown items. They may see another mushroom in the market, but rarely buy it. *Jongoler athoba amader asepaser sob jinis khete nai, kunta khawa jai or kunta khawa jaina bujte hobe. Jongle anek bishakto mushroom ase* (Bengali). [The Manipuri do not eat whatever they find in the local garden and forest. They know which is edible or poisonous because of their experiences and observations, or the Indigenous knowledge. For example, they know certain wild mushrooms in the forest are deadly].

The Owl and Natural Pesticides

The owl controls the insects on the crops' land. The birds maintain local ecology. The Manipuri community believes the birds are changing their habitats or facing some challenges in the area because birds are not seen regularly as before. *Amader bagane amra sobsomoy kaktarua rakhtam pecha jate boste pare. Akhon are pecha dekha jaina ager moton. Sokal bikal pakhi bosto are puka makor kheto* (Bengali). [The Manipuri uses *kaktarua* (scarecrow) in local gardens and farmland. The scarecrow is one effigy where birds sit and eat insects in the garden and rice fields].

Local birds such as the *drongo* (short-legged blackbird) and owl are regular visitors to this *kaktarua*, which looks like a dummy man. It is made of straw, broken clay pots, and used clothes. The *drong* sits there during the daytime to eat insects, whereas the owls sit at night to eat insects. The number of insects has increased significantly because owls are not seen as often as before. The community worries that owls are one of the species experiencing extinction in the area.

Climate Change and Impacts on Water

Climate change affects weather pattern

Bangladesh has six seasons: summer, rainy season, autumn, late autumn, winter, and spring. Ecosystem changes in the Sylhet region are due to the cutting down of trees and the destruction of the forest. For example, real estate developers, brickmakers, and loggers are responsible for cutting trees and *tila* (hillocks) in the Sylhet region. These trees control water runoff, water speed, and soil erosion. Siltation occurs in waterbodies with eroded soil. Moreover, the water catchment area is shrinking due to siltation. Sylhet was known to be full of natural creeks. These creeks have filled up or been encroached by land grabbers. The natural flow of water has been stopped. These creeks are connected and merged with the Surma River. Creeks worked like a natural drainage system in Sylhet City for several years. But now, flash floods have become a regular phenomenon in the Sylhet region (Islam et al., 2006, p. 303). Another critical point is that during the monsoon the Surma River suddenly becomes full of water and spills over nearby and inundates Sylhet City. The river cannot hold the water because of high siltation. It creates a barrier to natural water flow and reservoirs. Flooding has become a regular phenomenon in Sylhet City. Sylhet City was under water in May 2022, and the city was the worst affected ever over the past 100 years. In recent years, the rainfall in the monsoon has started to decline moderately (Akter et al., 2019, p. 7). "The impact of climate change on Indigenous communities is multifold, such as the advent of livelihood problems emanating from the drying up of streams and wells, groundwater depletion, depletion of wildlife, infertility of crops, and the mortality of seedlings" (Mustajib, 2020).

The community observes irregular climate behavior in the Sylhet region. One respondent stated, *Akhon gororomer somoy anek gorom ar bristir somoy ati bristi. Age ai rokom hoitona abong shilo-na* (Bengali). [Irregular rainfall is occurring during the monsoon season, and temperatures are increasing during the summer]. For example, the Manipuri have experienced heavy rainfall or no rainfall during the rainy season. Sylhet City is located on the bank of the Surma River. There is no water most of the time during both rainy and winter seasons. Moreover, dunes can be seen in the middle of the mighty Surma River. City dwellers of Sylhet experienced 38°C temperature in 2023, along with more humidity. However, they never had this type of heatwave over the past 35/40 years.

Drought

Climate change invites drought in the Kamalganj area. There was no regular season in the area that the community used to have 20-30 years back. Farmers used to follow the local seasonal calendar, which has changed over the years. For example, the rainy and summer seasons have changed.

> *Akhon season paribartan hoye gese, thik somoye bristi porena koyek boshor dore. Fole, adhunik panir pump bebohar karte hoi pani tante, sar lage, kitnashok kinte hoi. Sobai adhunik machine bebohar karte parena karon anek taka lage* (Bengali). [There has been less rainfall during the seedling processing period over the periods. So, the farmers are required to use a water pump for irrigation. Thus, farmers increasingly depend on mechanical farming techniques such as modern irrigation systems, fertilizer, and machine harvesting].

Farming costs have increased because of irrigation, pesticide procurement, and machine hiring. Farmers require irrigation because the water level has dropped. Some wealthy farmers use deep tube wells that further push the ground-level water down.

Brickfield Impacts on Communities

Brickfields bring air pollution and disease to the Manipuri area. For example, there is a large brickfield owned by a local politician near a Manipuri village (Ganga-nagar). The respondent Pari explained his observation and experience in the following way,

> *Mi mayamdo uraga nungaite. Yamna tonna leiba gumna uba fangli. Mi singdo masasu faba mande, brickfield-da maram oiduna. Thongan-da brickfield leire, mafam ado ufun kare. Maram aduna uhei, wahei, pambi parang, lousing hingbada ware* (Manipuri). [I felt sorry when I saw the

> local Manipuri's situation and the brickfield's impact on the area. For example, they are not healthy. The area is entirely covered with dust, covering trees, rice fields, and vegetables. The community is helpless to this brickfield and its owner].

The brickfield owners take firewood to continue their brick business. Therefore, wood cutting continues in the area of the Rajkandi reserved forest near the brickfield. The villagers suffer from respiratory and other diseases, but the brickfield continues to burn wood and release CO_2. As a result, the area has become environmentally polluted, and this affects agriculture, soil fertility, air, water, and people's health. The brickfield also changes the total environment of the area such as local flora and fauna. The Manipuri have yet to complain to the authorities about stopping the brickfield because the community cannot face reprisal from the owner of the brickfield.

Mango Plucking and Displacement

This interview was conducted over the phone on April 9, 2024. We chose a community Elder as a key respondent because he is an outspoken social activist. Ten Manipuri families lived in a small village Chingtham Leikai. Local non-Manipuri plucked the mangoes without consent from the Manipuri community. This action hurt the feelings of the community and was against their culture. One of the Manipuri boys protested, but he was beaten severely by local goons. After a few days, the same things happened again in the area. Then, a Manipuri elder (woman) was protesting the mango plucking by non-Manipuri without any permission. This time, she was attacked and received a head injury from a brick. The Manipuri filed a complaint to the local union parishad (UP) chairman but did not get any proper support. The local Manipuri community felt threatened because they had less manpower and were economically weak. The community thought that the non-Manipuri were playing some tricks, so they left the area. Within the Manipuri community there was mass displacement between 1985 and 1992. The Manipuri cannot imagine being without land. Pari mentioned this in the following way, *Hainou hekpa namfuda asisu khudam amani akhoina karisu touba ngamloi haina* (Bangali). [Forcibly plucking mango from the garden is a message that the community is helpless and vulnerable]. *Yu thakna leiraga mina usit-kadara. Leirik haiba watpa asinasu mina thou sabidare* [If you are drunk, nobody will respect you. If you are educated, people will respect you].

Alcohol consumption is an issue in the Manipuri area. The community believes that alcohol makes people lazy and weak. For example, work culture has diminished after alcohol came to the area. Moreover, alcohol brings poverty and disease. This alcohol is locally made in the tea gardens, which do not maintain any quality. Thus, the number of widows in the area has increased significantly. Moreover, non-Manipuri also do not respect a person who is an alcoholic.

Challenges of the Manipuri Exacerbated by Climate Change

The increase in the number of insects and reduction in owls and birds recounted earlier has resulted in the increased use of pesticides in the area by agricultural producers. A new cost has been added to farmers. Moreover, pesticides are harmful to human health. However, the community is indirectly consuming chemicals through the food chain. Indigenous Manipuri elders noted how these pesticides also kill important *kechu* (worms) and insects, which are beneficial for soil fertilization. For example, worms retract soil from the ground to the sunlight and make the soil fertile. They contribute to increasing crop production yield. *Ager krishok ato parimane sar abong kitnashok bebohar karten-na akhontar tulonai* (Bengali). [Earlier, the farmers did not use such chemical fertilizers and pesticides].

Most of the farmers use modern pesticides instead of earlier traditional pesticides. Previously, the farmers used ashes instead of chemical pesticides to control insects. Ashes work like pesticides because they kill insects. However, it becomes difficult to kill the insects with traditional pesticides due to the increasing number of insects.

Religious Practice and Challenges

Every Manipuri household has a place for Sanamahi worshipping. Sanamahi is the supreme god of the Manipuri. A Manipuri is required to worship twice a day: morning and evening. Manipuri worship *numit* (the Sun) in the morning; their Sanamahi practices dictate that they get up before the sun rises. However, the younger generations show less interest in following this ancient practice. Therefore, many times it is only the Manipuri elders who are following the sun worshipping. Worshipping the sun means there is a connection between nature and the community. For example, Manipuri days are being counted based on the sun. The examples are given in Table 3.1.

Akhonkar projonmo digital divice niye besto thake. Bap-dadar gyn dore rakhar moton mansik tader nai (Bengali). [Today's generations prefer to be busy with digital devices such as mobile phones and computers, but they are reluctant to preserve and practice Indigenous knowledge and the Indigenous practices of counting days based on the sun.]

Table 3.1 Manipuri and English language of days

Days (English)	*Days (Manipuri)*
Saturday	*Thangja numit*
Sunday	*Nongmaijing numit*
Monday	*Ningthoukaba numit*
Tuesday	*Leibak-pokpa numit*
Wednesday	*Yumsakeisa numit*
Thursday	*Thagolsel numit*
Friday	*Irai numit*

The younger generation is reluctant to learn Indigenous knowledge from elders. The younger generation focuses more on electronic devices than learning Indigenous knowledge from Manipuri elders. For example, most of the younger Manipuri have yet to learn about rice plantation ceremonies conducted in the field concerning nature and earth. Still, this initial rice plantation ceremony is practiced among the community by the elders. However, it is under threat because of a lack of preservation and practices. One respondent claimed that he has some Indigenous knowledge. For example, his father knows about the seed selection and preservation for the next plantation. This community still applies Indigenous knowledge to select the seed. However, the next generation may not have this Indigenous knowledge. One respondent has a vision,

> *Ami akta model Manipuri para dekhte chai Sylhet city-te. Ami Mayor ar sathe dekho korobo abong proposal dib unake- kibhabe Shibgonj Manipuri parake model hisabe preserve karte pari* (Bengali). I wanted to see a model Manipuri para (village) in Sylhet city. I will meet with the City Mayor soon to propose if Shibgonj can be made a model Indigenous Manipuri para.

This key respondent suggests a model Manipuri para in Sylhet City to retain the Manipuri identity in Bangladesh. The respondent is also worried about whether the community can maintain its identity in an urban area like Sylhet. The respondent has a development organization in Sylhet. He is aware of the changes in the Manipuri community. According to the respondent, if the Manipuri para model is developed with the help of the Bangladesh government, it could contribute to the socioeconomic development of urban Manipuri in Sylhet.

Discussion from Learning Reflections

The Manipuri has experienced climate change, such as irregular rainfall and drought. Deforestation is one of the reasons for less rainfall in the region. Manipuri and mainstream communities face similar water crises during the monsoon. Regular rainfall helps preserve the production of local vegetation and the environment in the area. However, the Manipuri cannot start seedling, planting, cultivating, and harvesting based on the season anymore. The local seasonal calendar has changed. The Manipuri have Indigenous knowledge of adapting to climate change because they have been facing it for years. Similarly, Indigenous people of Chittagong Hill Tracts (CHTs) consider climate change natural and adapt to their local knowledge (Garai et al., 2022).

Brickfield uses massive amounts of wood from the local forest. Therefore, brickfields contribute to deforestation in the Sylhet region. Deforestation affects the people's livelihoods, health, and environment. Moreover, deforestation affects local flora and fauna in the area. Deforestation minimizes groundwater infiltration, recharging the water, and this deficit creates droughts (Islam et al., 2006). Influential people also cut hillocks to collect soil for landfills and brickmaking. They are influential, so ordinary people rarely make complaints against them. Some of them have connections with the local political elite, and in our interview one was a local politician. The Ministry of Environment and Forest of the Bangladesh government could take decisive action against the brickfield owners, particularly those operating near the Indigenous community, such as Manipuri and CHTs. Recently, the Minister of Environment, Forest and Climate Change has passed ordinances such that Brickfields stop operations located near Dhaka (Mahmud, 2024).

According to their culture, the Manipuri must have an *ingkhol* (garden) where they plant traditional trees, vegetables, and herbs. For example, they plant *yongchak*, yam, neem tree, awa-phantom (false coriander), and *mayangba* (hoary basil) in their sinkhole (home garden). These are traditional ingredients for cooking and medicinal purposes. The Manipuri community has been using these from generation to generation. However, the community observes that some significant trees and herbs, such as *yongchak*, are dying slowly. Most houses have vegetable gardens that produce vegetables for their personal needs (Ahmmed & Singh, 2007, p. 63), but for some this way of life is being lost.

The Manipuri believe in *umanglai* (forest god). The forest god–related ritual is conducted during the Manipuri New Year (*Shajibu Cheiraoba*) and *Lai-haraoba* festival (the largest religious festival of the Manipuri). The Manipuri belief and culture instruct that there must be a small forest nearby the house, which is their sacred grove. These sacred groves maintain soil firmness and avert soil erosion (Singh, 2009, p. 125). There is a religious, spiritual, and cultural connection between the local forest and the community. Sometimes, the Manipuri gardens are considered the forest. The Manipuri home garden is similar to a mini forest because it covers ritual and medicinal trees. These trees maintain local ecological balance. This Manipuri *umanglai* belief system has a deep idea of conserving plants and animals, according to Malhotra (cited in Singh, 2009, p. 120). For example, birds use the trees for making nests and as a breeding place. Manipuri elders predict weather forecast based on birds, reptiles, flowers, and other local insects. For example, when birds are chirping and behaving in an abnormal way, the community believes that there must be a snake in the area. Moreover, early blooming of a flower and fruit tree indicates local climatic behavior and changes. Another example is that when the sky is dark on the north side, the rain will come soon, according to the traditional Manipuri belief system. Those who are living in urban areas become challenged to maintain their forest because the land is shrinking or sold out to non-natives. Thus, development and land grabbing impacts the Manipuri culture and identity.

The Manipuri is semi-vegetarian because they eat fish. They do not use onions and garlic. However, they widely use ginger, turmeric, and local herbs as cooking ingredients. The community also changes food habits because medicinal plants and herbs face extinction due to regional climate change. For example, they have started using processed ingredients from the market, which makes their lives difficult. The items are expensive in the market. The modern market products make them more vulnerable to increasing prices and food insecurity. The Manipuri is an underprivileged community. They need to be stronger socioeconomically and politically.

> Their sociocultural settings are inevitably intertwined with the nature they live in. They used to engage in agro-forest farming systems for generations. This process is inextricably related to seasonal patterns. But, this nexus between nature and community is being adversely affected due to climate change implications.
>
> (Mustajib, 2020, n.d.)

The Manipuri eat *charuyen* and *uyen* as part of the vegetable item. The research suggests that these items are not available nowadays. A study established that macro-fungal diversity in the Indo-Bangladesh region is threatened due to the effects of climate change (Bhattacharjee et al., 2015, p. 75).

Yongchak trees are facing extinction in Manipuri areas in Kamalganj. One of the key respondents claims groundwater depletion may be responsible for the disappearance of *yongchak*. Respondents also argue that mobile and mobile towers may be another reason for the *yongchak* trees dying in the area. Similarly, a Manipuri writer points out that young trees have had difficulty surviving since the mobile tower came to Manipur (Singh, 2020). Therefore, the *yongchak* tree faces extinction in the Kamalganj area due to the mobile tower and its invisible wave generation.

Herbal Trees and Challenge

This research has identified that important herbal trees and herbs are disappearing slowly from the area. However, they play a crucial role in the cultural and health maintenance of the Manipuri. The Manipuri have a record of using herbal medicine. Local trees have a deep connection to local people and their culture. For example, they use the neem tree and its leaves for medicinal purposes. Most of Manipuri's houses plant neem trees. It is also a mandatory food item during Manipuri *Shajibu Cheiraoba* (Manipuri New Year). As well, neem is used during culturally related lunches such as *Tarpon* (food offering to dead ancestors) and other lunch programs. The Manipuri also use neem sticks as a toothbrush instead of a modern toothbrush. The branches are made into small pieces to make it usable. However, this tendency has been replaced by modern toothpaste. An important observation is that the Manipuri do not eat neem at night. The community believes it creates toxins at night. Moreover, the community uses neem as a natural pesticide tree. They also think that neem trees can drive pestilence and disease from the area. The Manipuri have been using neem for ages (as long as they can remember). Neem trees have been affected less than the *yongchak* tree in the area.

Indigenous Knowledge

Indigenous knowledge of the Manipuri contributes to adaptation to climate change. The Manipuri carry some Indigenous knowledge to adapt to climate change. For example, they strictly maintain the traditional Manipuri housing structure because it is an eco-friendly housing design. The housing design considers the earth, universe, and other living things in the area. Manipuri housing design reflects their culture and how they adapt to the local climate. "Manipuri vernacular houses give due respect to passive and natural environment control measures in buildings. Construction techniques and material culture show a suitable response to the microclimate of the Sylhet region" (Saha & Chowdhury, 2021, p. 104). However, modern housing design is now impacting Manipuri housing design. As a result, these new housing practices affect Manipuri culture, local survival techniques, and environmental connections in a negative way disconnecting the Manipuri from their natural environment and the earth.

The windows are placed in such a way that cool breeze air comes from the south window into the home and can be released through the north window. Manipuri's housing design also instructs the planting of trees surrounding the house while maintaining an elephant-sized distance (Singha 2015). The Manipuri also believe that trees can protect them during the cyclone. Similarly, the Indigenous community of CHTs plant trees surrounding their houses to protect them from speedy wind and storms. Moreover, they also cut the larger size of branches to save lives (Garai et al., 2022, p. 8).

Indigenous knowledge has been an essential part of protecting the environment in Bangladesh. Indigenous elders of CHTs possess valuable Indigenous knowledge and wisdom as an essential part of the culture to adapt to environmental degradation (Garai et al., 2022, p. 8). Datta (2018b) claims that elders and knowledge holders should be included as an essential part of community research because it is based on Indigenous knowledge. A "two-eye" approach that can develop a co-learning and relationship-building between Western and Indigenous knowledge can pave the way for learning and coexisting respectfully (Galway et al., 2022, p. 9). Corntassel and Hardbarger (2019) suggest focusing on local stories and knowledge, language, Indigenous land, and community to form a foundation for Indigenous nationhood. CBA should not be isolated activities within a community (Kirkby et al., 2015). This study further argues that CBA needs to be scaled up and scaled out. Scale-up helps to connect local needs to upper management. Then, scaled-out contributions make it a larger project from a small project. CBA requires inclusive, empowering, and context-specific support for the adaptation needs of those most vulnerable to climate change (p. 2).

Land and Identity

Land is an essential part of identity maintenance among Indigenous people in Bangladesh. For example, the Manipuri identity has developed and retained a land-based culture. Key respondents suggest that a Manipuri model para is required to preserve Manipuri culture and total identity in Sylhet. Cultural and religious activities are performed on land. Thus, a Maniprui model para can protect the urban Manipuri in Sylhet. For example, there were Manipuri in Dhaka. The place is known as Manipuri para in the Tejgaon region in Dhaka; however, currently, only a few Manipuri are living there. Urban Manipuri in Sylhet may face similar consequences if they do not take proper steps to protect their land, identity, and culture. For example, "the connection to IEKK land-based stories is crucial for understanding and practicing environmental sustainability and for the wellbeing of Indigenous and non-Indigenous peoples" (Datta, 2023, p. 4). However, there have been complaints and reports that Indigenous lands have been encroached on and taken over by corporations, local politicians, and tea estate companies in Bangladesh (The Indigenous World 2023, 2023, pp. 162–164). This is what the Manipuri believe happened to the Manipuri para in Dhaka.

The daily activities and the culture of Manipuri are interconnected with the land. Sometimes, the Manipuri face displacement from their ancestral territory. Many Indigenous communities face displacement and exclusion from land as a common predicament (Uddin, 2014, p. 1). For example, government and commercial agencies continue land dispossession from customary land rights that contribute to the suffering of Indigenous communities in CHTs (Uddin 2014, p. 3).

The key respondent Pari mentions that the relationship between younger and older generations has changed. The older generation of non-Manipuri were more kind to the Manipuri. However, new generations are aggressive and have less respect for the Manipuri. When the community feels threatened, they leave the place. Non-Manipuri buy the land below the market price. Sometimes, the seller never gets the money. In fact, mango plucking is an example signaling to the Manipuri that their land has become a target for landgrabbers. Landgrabbers create homelessness and an identity crisis among the Manipuri. Sometimes, the local administration is reluctant to take action against the perpetrator. This reluctance encourages the landgrabbers and other goons to repeat the crime against the community. Other academics have identified the colonizer as the one who aims to get Indigenous lands and knowledge structures disrupted and thereby augment an affinity with dominant politico-economic institutions, thereby coercing communities to disband (Nightingale & Richmond, 2022, p. 7). Sometimes, the authorities tried to apply a "green grabs" policy to the land of the Indigenous in 2000, such as creating the Madhobkundo eco-park in Bangladesh. By creating a park open to all from the Indigenous peoples' land in which they are living, they effectively denigrate the Indigenous community, its culture, and its practices. However, the government changed the policy while respecting the Khasi community residing in the park. Landscape perspective and affected people must be included as co-producers of knowledge to solve land-related issues (Hunsberger et al., 2017, pp. 306–318). The study further suggests that collaborative, engaging methods will contribute to sustainable solutions in relation to climate mitigation and land-dealing based on Myanmar and Cambodian experience (p. 318).

Conclusion

This chapter reveals that there have been climate variations compared to earlier periods in the Sylhet region, documenting the lived reality of the Manipuri people and their experience of climate change. Climate change impacts the total Sylhet region but inordinately impacts the Manipuri's livelihoods, local environment, and ecology. Important birds are decreasing while the community is increasing the use of chemical fertilizers. Then, the community suspects the mobile tower is responsible for dying of the bean tree, *yongchak*. After the mobile towers were set up in the Manipuri area, the traditional food *yongchak* almost disappeared. Manipuri food sources are vital in keeping their bodies fit over the generations. However, Indigenous food extinction becomes a threat to their healthy lifestyle and culture. The Manipuri apply Indigenous knowledge to nurture and maintain the local herbs and medicinal vegetable items. Moreover, the Manipuri have a culture of keeping a *ingkhol* (garden) for food sources such as herbs, vegetable items, and fruit-producing trees. Manipuri culture, beliefs, customs, and housing structure are embedded in the environment. The land is an integral part of Manipuri's identity and culture.

The chapter suggests that ancestral land should be protected and preserved to retain the Manipuri identity and allow them to adapt to climate change. Indigenous identity, culture, and livelihoods have been embedded in the land. Landlessness will lead to the identity crisis of the Manipuri like many other Indigenous communities of Bangladesh. The Bangladesh government should take proper steps against landgrabbers and those who destroy the environment in the name of business expansion in the Indigenous area. Creating a Manipura para model in Sylhet would advance community adaptation and stem the increasing vulnerability of this Indigenous population. Indigenous voices should be listened to, respected, and included by decision-makers while making development and policy decisions (Datta, 2024b, p. 8).

This research has a limitation in finding storytellers. The number of Manipuri storytellers is declining because of modernization and the lack of preservation of Indigenous knowledge. This indicates that retention of Indigenous knowledge is a challenge among the Manipuri. Indigenous knowledge is being lost in both urban and rural areas. How the *youngchak* tree disappears and how mobile towers and their waves are interconnected could be potential research topics in Bangladesh because they are essential food sources for Manipuri. The preservation and archiving of the Indigenous knowledge of the Manipuri would also be another worthwhile research area.

4 Flood Disaster Impact and Adaptation

Learning Reflections from Marma Indigenous Communities, Chittagong Hill Tracts in Bangladesh

This chapter examines the impact of climate change on the Marma Indigenous community in Bandarban, located in Bangladesh's vulnerable southern region. The area faces significant threats from human-created environmental disasters, such as flash floods, droughts, and landslides, exacerbated by climate change. The August 2023 flood serves as a case study, highlighting the severe consequences for the Marma community, who rely on the region's natural resources for their livelihood. Through stories from community elders and women, the chapter explores the causes of the flood, its impacts, and the coping mechanisms and adaptation strategies employed by the Marma community. The chapter concludes with a reflection on the challenges faced by the Indigenous community and recommendations for future resilience efforts.

Bangladesh's southern region is highly susceptible to human-created environmental disasters, with communities relying on forests and natural resources at high risk (Islam et al., 2015). Climate change may further exacerbate the precarious conditions of tribal groups in southern Bangladesh, as they live on marginal land and rely on forests for subsistence (IPCC, 2007). The weather is changing. These changes may be indicated by variations in the patterns of rainfall, droughts during rainy seasons, late monsoons, recurrent floods, and mild winters. It is commonly acknowledged that these changes in the weather have had a major impact on southern Bangladesh, especially in the southwest of the country (Fardous, 2011). For example, climate change–related disasters are expected to have a severe impact on the Chittagong Hill Tracts (CHTs), as evidenced by the devastating flood that occurred in August 2023. Heavy rains on August 7, 2023, caused major rivers in southeastern Bangladesh to surge well above their dangerous levels, resulting in flash floods that left several areas in Chattogram, Cox's Bazar, Bandarban, and Feni inundated (ADRC, August 7, 2023).

Climate change is posing several issues for CHTs. Recurring climate-related dangers include flash floods, droughts that cause water scarcity, landslides caused by prolonged rainfall, the drying up of water bodies from a lack of rain, and soil erosion. Due to their extreme poverty and isolated rural location, the majority of the local community is being significantly impacted by these dangers, negatively affecting their lives and means of survival (Haque et al., 2021a). Although soil erosion and deforestation caused by humans are the primary

DOI: 10.4324/9781003536345-4

causes of flash floods and landslides in the CHTs, it is likely that precipitation and storm surges brought on by climate change will cause a significant increase in both the frequency and severity of these human-created environmental disasters (Gunter et al., 2008). Landslides are common but were particularly severe in 2007, 2010, and 2017. In 2007, 135 people were killed, and 1.5 million people were affected by heavy monsoon rainfall. In 2021, 30,384 people were affected across Bandarban District, with abnormal precipitation causing widespread destruction (BBS, 2022; Haque et al., 2021a; Ullah, 2024). Climate change affects everyone equally, but because of their closer proximity to the natural environment and their early experience of recognizing and adjusting to changes related to climate change, Indigenous people in CHT, Bangladesh, will be particularly vulnerable (Haque et al., 2022).

The government of Bangladesh acknowledges 12 ethnic groups (i.e., Indigenous communities) that call the CHTs home. These are Mrus/Mro (Moorangs), Marma (Maghs), Bawn/Bawm, Chak, Chakma, Khyang, Khumi, Lushei/Lushai, Pangkhua/Panghua, Tangchangya, Tipperas (Tipras), and Bengalis (Ministry of Health and Family Welfare, 2016). People from all community live in Bandarban City. We visited the bank of the Sangu River. That is the place where the Marma, Barua, Chakma, and some Khyeng community people live along with the mainstream. They all have suffered the impact of the flood. The effects of the flood can be significantly reduced by establishing good resilience and scientific knowledge about the disasters. However, the literature does not identify any such support and resilience for the Indigenous communities in Bandarban District. The Indigenous community all over the world is so vulnerable (Axelsson & Sköld, 2006). Food poverty, landlessness, and illiteracy are the indicators of vulnerability among Indigenous people as well as non-Indigenous people (Gunter et al., 2008). It could be caused by cultural differences, unequal access to resources, and the marginalization that is generally connected to racial injustices in social, economic, and political situations (Akinola et al., 2019; Wigtil et al., 2016). This study will discuss the flood that happened in August 2023 in Bandarban, its impact, and how the Marma Indigenous community survived during and after the flood.

This chapter aims to achieve three primary objectives: (1) investigate the reason for the flood, (2) find out the impact of the flood, and (3) learn about the coping mechanism and future adaptation (reality and expectations) strategies of the Marma Indigenous community in Bandarban. The stories here presented came from the Marma Indigenous community. We have listened to the stories of the elderly members. We have presented our data analysis results where we address the issues, challenges, and recommendations that were shared by the Indigenous community members. We have concluded our study by including our learning reflections from the Marma Indigenous community.

Researcher Positionality

The role of researcher positionality is of utmost importance in Indigenous research, as it significantly impacts the research process and raises ethical considerations (Datta, 2018; Wilson, 2008). It is essential to recognize that Indigenous research frequently delves into sensitive subjects and engages with marginalized communities. Researcher positionality encompasses their cultural background, personal experiences, and social identity, all of which can profoundly influence their interactions with Indigenous participants and their comprehension of the issues at hand. Acknowledging and understanding one's positionality is crucial for approaching research with humility and respect, reducing the risk of perpetuating harmful stereotypes or colonial attitudes inadvertently. It empowers researchers to establish trust within Indigenous communities, nurture meaningful collaborations, and formulate research questions that are culturally sensitive and aligned with Indigenous worldviews.

I (author-4) am a member of an Indigenous community in Bangladesh and also belong to one of the minority communities of the country. My involvement in this study stems from a desire to give voice to the unheard. With enthusiasm, I listen to the stories and aim to become a storyteller on behalf of Indigenous communities to the wider world. My affinity with other communities motivates me to collaborate with them. From the stories of the Indigenous communities, I got all the motivation and strength to work with them. They empower me, educate me, and enhance my intentions.

In addition, our positionality helps to address historical power disparities in Indigenous research. Instead of being active participants or collaborators, Indigenous communities have frequently been the focus of research. We were able to redefine our research agenda by making sure that the research serves the needs of the community by critically analyzing our positionality. Establishing ethical, respectful, and empowering research relationships with Indigenous communities and advancing knowledge that truly serves their interests and well-being are contingent upon our positionality.

Theoretical Framework and Methodology

The goal of a decolonial research framework is to confront and address the lasting effects of colonialism in academia and beyond through a critical and transformational approach to study (Datta, 2018b). It is acknowledged by this paradigm that colonialism had a significant impact on historical context as well as research techniques, viewpoints, and epistemologies. Decolonial studies aim to dismantle and undermine prevailing narratives and systems that have historically disadvantaged Indigenous people as well as other oppressed groups. The statement acknowledges that customary research methods frequently sustain disparities in power by objectifying and taking advantage of Indigenous knowledge and communities. It acknowledges that conventional research methods frequently uphold power disparities by objectifying and taking advantage of Indigenous knowledge and peoples. Therefore, a decolonial research paradigm critically examines how colonialism has shaped the research process while putting Indigenous perspectives, knowledge systems, and worldviews front and center.

The main goal of a decolonial research framework is to build fair and respectful collaborations with oppressed groups, including Indigenous people. This strategy respects these communities' sovereignty and self-determination by giving them the freedom to choose the study priorities and methods. It also highlights how crucial it is for researchers to actively seek to reduce damage and advance ethical research techniques while also admitting their own positionality and prejudices. A decolonial research framework is essentially an appeal to reconsider and alter the research environment with the goal of achieving more inclusive, egalitarian, and decolonized knowledge creation that honors the various ways that marginalized and Indigenous groups know and are known.

We used decolonial storytelling approaches and deep listening in line with the decolonial research methodology. We followed customs and rites that are Indigenous to the area and learned stories from the Marma community who lived in cities in August 2023. One of the most important tactics for showing regard and respect for the viewpoints of Indigenous communities was deep listening. Storytelling is essential to reestablishing a connection with land-based learning since Indigenous stories are intrinsically linked to the land and are defined by the land, language, and the spirit of the people (Datta, 2018). We adhered to established institutional research procedures as well as customary Indigenous rituals and ceremonies.

Findings from Learning Reflections

The study was conducted in Bandarban since this area has the largest Marma population density. The researchers (Authors 1 and 4) visited a few areas where all the Indigenous people live; Ujani Para is one of them. In that area, people suffered a lot because they lived by the river. When a flash flood came, every single person had to suffer the consequences of the flood.

We listened to the recorded stories several times, concentrating on different facets of them. We concentrated on the main subjects and related issues that we have identified. After sharing our stories and giving the recordings another listen, each of us took some time to reflect on what we had learned. We were able to learn with respect thanks to our own reflections. Our learning stories featured emotional and often challenging accounts from Indigenous women about their opinions and experiences. The choice of quotes and anecdotes to utilize is often influenced by personal preferences. The majority of the lines that we chose were memorable, thought-provoking, and important enough for us to consider and oblige to memory. We continuously engaged in listening and reflective learning as we created the primary themes and sub-themes (Figure 4.1). How the Marma Indigenous community is affected by climate change, especially the flood, and then adaptation tactics are the main concerns. Then we focused on the adaptation that Indigenous community people provided.

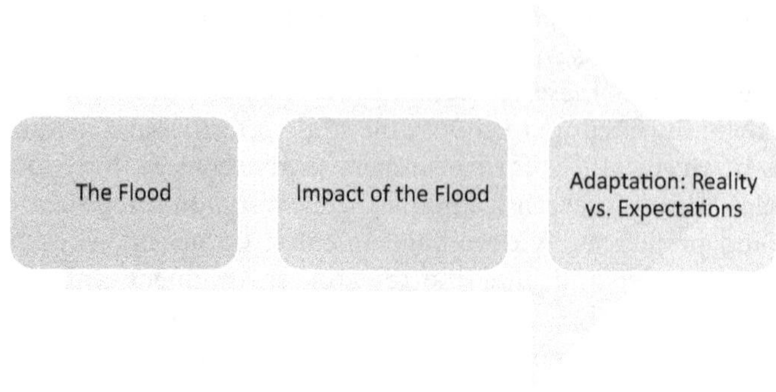

Figure 4.1 Main Themes and Subthemes of the Results.

Theme 1: The Flood Disaster

In this part, we will discuss the types of flood, the reasons behind the flood, and warnings about the flood. In the year 2023, there was a massive flood in Bandarban. The flood made the city shut down for 20 days, and most of the city went under water. The Indigenous man 3 said, "There was a flood in 1997, which was similar to this one. After that, for a long time we haven't had any major floods. This time the flood was much worse." A flood or any disaster is not so usual in that city. Also, when the flood happened, it lasted for a long time, and the inhabitants could not imagine it. The Marma man said, "The flood stayed for 20 days. This time we could not eat. They will not provide anything if we are not relatives of those people. They have no mercy." For a long time, the Indigenous people had a painful life. There was lack of food, drinking water, and medicine,. and the memory of the flood was so horrible that they still haven't forgotten about the flood. A Marma Indigenous woman said, "We are afraid of the flood. It was one year ago but the fear exists. If it happens again, we will die." The Indigenous people have expressed their fear about the disaster. They know that if another disaster like a flood comes they will not survive.

Reason

We tried to find out the reason for the flood. We listened to what the Indigenous community people had to share. The Indigenous woman 1 said, "Maybe this time the rain was much heavier. When the upper areas face heavy rain, more water flows here. We think, if heavy rain happens in Thanchi and Ruma then this type of flood can occur." Some people of the Marma community believe that the flood was created by the heavy rainfall to the upper hilly areas of Bandarban District, whereas other people believe that the city has become an overpopulated one and new houses are built by seizing the river. The Indigenous man 1 said, "We have only one river for water circulation-Sangu, which is a small river. The river has been captured because of new establishments and population influx, so it can be the reason behind the flood." In Bangladesh, this is very common for settlers to grab the land of the Indigenous community people. So, this is also true that the river and river bank are also being grabbed by the mainstream settlers. By doing this, they are making the Indigenous people suffer. These Indigenous people are so poor and vulnerable, and the disasters that are happening lately make them suffer out of limit. They are frustrated, and they have nothing to do about it except sharing their sorrow and sufferings. Indigenous man 2 said, "It is God's wish. People are not righteous and religious now. (Maybe he was trying to say that we people have gone rough so God is punishing us by floods). People became sinners, they didn't obey religions. That's why floods happen." The Indigenous community people can do nothing except pray and think about the justice of nature. He also added, "People are also helpless. People don't respect elderly people nowadays." So, these people also think that the disasters happen because of the wrong doing of people. People are doing wrong to nature, and nature is giving back to them what it gets from them.

Type of the Flood

We wanted to know the type of flood from the Marma Indigenous communities. They have described the flood as very dangerous and devastating. The Marma Indigenous woman said, "We've seen floods, but not in this magnitude. Our house never got broken in the flood. Not even in the last 3–5 years." So, this year the flood was so extreme that their houses were severely damaged because of the flood. Also, the impact of the flood was so troublesome. The Indigenous woman said, "We have never faced this much disaster and trouble in our entire life. We cannot even count the loss." The Indigenous people suffered a lot from the flood. According to them, this flood was the most extreme flood they had ever seen and the loss was also unexplainable. We listened about the flow of water in the Sangu River during the flood. Indigenous woman 2 said, "This time water flow was so intense that everything got washed away." The Indigenous people living near the Sangu River suffered a lot due to the flood of 2022, and they lost their utensils and belongings in the flood. When a flood hits an area, the people become vulnerable and after the flood they rush to return their houses and try to fix them. This time they lost everything. This flood was so extreme that they could not start the process instantly. Indigenous woman 3 said, "Usually whenever floods happened here that was bearable. We used to clean our houses after the flood water was gone. But this time water flow was so intense that everything got washed away." Even after the flood had gone, they could not return to their everyday normal life.

Warning or Prediction

We asked the Indigenous community people whether they got any disaster warning. Indigenous woman 1 said, "No one can predict when any natural calamity will occur. It can happen anytime." So, the community people think that disaster can happen anytime and nobody can predict it. Indigenous woman 3 said, "The authority provided signals of heavy rain. We have seen this kind of warning every year. But never seen this much water. And it happened all of a sudden." In that area every year the Union Parishad office provides the forecast or warning of extreme weather, such as heavy rain and extreme heat. But, the community took the signals lightly because they had never experienced anything massive like the flood. Marma Indigenous woman 2 made this clear saying, "We have never seen something like this, so couldn't even think about this kind of destruction. The flood was so sudden that we couldn't move out our belongings." It is clear that even if they were warned about heavy rain, they could not take it seriously. Because nobody could think that they would face something like this in that area. She also added, "As we live in the lower side of the area, we were affected first and couldn't take any precaution. People from higher lands got some time to move out their belongings. But we did not get the time." Indigenous people have their knowledge and prediction for weather, which is most of the time well predicted. It is certain that something must have changed that an area that was never affected by any human-created environmental disaster is now a place of extreme disaster. It is obvious that there is room for giving attention to this perspective.

Theme 2: Impact of the Flood

In this part, we will discuss the impact of the flood. We found that communities' houses were destroyed, impacted their food, water, and health, and belongings were lost.

House Destroyed, They Lived

In 1997, Bandarban witnessed a devastating flood that took the lives of many Indigenous people. After 1997, people witnessed another devastating flood in 2022. The entire city was under water for 18 days. People didn't die this time, but they lost their home, their property, and everything they had. Comparing these two floods considering the damages, Indigenous woman 1 said, "I've seen people get washed away with the flood in 1997, this time no life was lost. But people lost their houses and property, livestock etc." Losing their homes, they have no option but to stay in a rented house. They do not even have the money to pay rent. They lost their homes, all their properties, and assets. Indigenous woman 2 mentioned, "My house got destroyed, and I'm compelled to stay in a rented house. I don't have money." They said they never faced a disaster like this one. They lived, but the damage was beyond repair. The Indigenous woman said, "We have never faced this much disaster and trouble in our entire life. We cannot even count the loss. Everything got flooded. Even the ceiling got drowned."

Damaged House Items

The damage of the flood was beyond repair. The loss was irreparable. The money they received in the name of assistance or help was nothing for the damages that occurred. Indigenous woman 1 said, "Houses got destroyed and all our belongings got washed away. But there has been help too, which was very little." Even the utensils of daily use got washed away in the flood. Indigenous woman 2 said, "Everything got washed away. Even the utensils of the house also got washed away by the flood." The Indigenous people didn't estimate the magnitude of the flood; they did not even think the water level would rise up to that level. People lived by taking shelter in higher areas. Indigenous woman 3 said, "People from higher areas sheltered us so we had to take shelter there. We didn't think that water would come this high."

They not only lost their houses but also their belongings, domestic animals, and poultry. This affected the Indigenous people economically as their incomes depend on what they breed. Regarding this, Indigenous woman 4 mentioned, "We had poultry (Hens, Duck). But all got washed away by the flood water." Grieving about their losses, Indigenous woman 5 claimed, "Our house got destroyed. Many things got lost with the flood water."

Food and Drinking Water Crisis

Water was everywhere, but not a drop was available to drink. The Sangu River is the only water source for more than 10 Indigenous communities. During the flood, there was a huge crisis of drinking water. There wasn't any supply of drinking water, nor were the communities given any purifying materials to purify the water. There was also a crisis of food; they starved during the flood. Indigenous woman 1 said, "There was huge difficulty with food and water." The responsible authorities did the least they could do and supplied the bare minimum. Sometimes, it's not even the bare minimum. Indigenous woman 2 talked about the amount of supply of food and water by the municipality during the crisis. "The municipality gave water twice and khichuri twice but it was not enough for all the affected people," said an Indigenous woman. The scarcity of food and water made them suffer, which is visible in the comment of the Indigenous woman, "There was a scarcity of food and water."

Impacts on Study

The flood affected the entire education system of the community. They already had only a few educational institutions; in the flood, they lost their books and educational instruments. The Indigenous people lost their homes and starved during the flood, so it was quite impossible to study as they didn't have any money to continue studying. Indigenous man 1 said, "Students have been affected massively. They've lost their home, educational instruments, books everything." After losing books, notebooks, or other educational instruments, it became very difficult for the students to go back to their school. They didn't have money to repurchase those items. Indigenous man 2 stated, "Students couldn't go to school. All their educational instruments like books and notebooks got washed away." As there are a very few schools, colleges, or educational institutions in the hill areas, many students come all the way to the hostels to continue their studies. They were tremendously affected by the flood. Indigenous man 3 said, "Many students come from afar, stay in a hostel and study. They've been affected a lot." The number of students dropping out escalated. "My grandchildren who were in the college, they had to leave the studies. They never started studying again," stated Indigenous man 4. The problem is, in this vicious cycle of poverty after the flood, the return of the students to the classroom is pretty much unsure.

Impacts on Health

The flood brought many health crises as well. The lack of shelter and a sanitation system made it impossible to maintain hygiene. There was no option to use a washroom so the Indigenous communities had to defecate where they stayed, which made the area unhealthy and contaminated. Indigenous woman 1 stated, "Everyone was there in the shelter. We had no option to use the washroom. We had to do where we stayed." While staying in an unhealthy environment with a lack of food and fresh water, people became sick. But they didn't even have enough money to buy medicines. Indigenous man 2 said, "We have faced some problems and had to buy medicine for that. It is also true that we had no money to buy medicines." The municipality didn't provide much for those who were sick. Indigenous man 3 said, "We have got fever. My wife got some medicines and a blanket from the aid provided by the municipality."

River Pollution

According to the Indigenous community Elders and Knowledge-keepers, the river has been polluted by the flood. The Sangu River had fresh water, but after the flood, the only water source for the Indigenous community got polluted. "The river was better in the past. But it is now polluted. After the flood, the river is in a bad condition now," Indigenous man 1 mentioned. This flood was so destructive that it caused river erosion. According to the Indigenous people, they've witnessed many floods, but they never witnessed this much severe river erosion. Indigenous man 2 said, "We have seen flood but we have never seen this kind of river erosion. Every day the bank of the river is getting broken." Many Indigenous people live on the bank of the river; if the erosion continues, many Indigenous people won't be able to stay in their homeland and will become homeless.

Theme 3: Indigenous Community-Led Adaptation

Before this flood disaster, the Indigenous people never had seen something like the flood because in the hills they do not witness floods. They have experienced landslides, droughts, extreme heat, etc. As they haven't seen flood and do not know what to do after the disaster, the Indigenous community people mentioned some aspects of what they need in life to survive after the disaster. They also told us the stories of their sufferings. While talking about the flood, they mentioned old-age allowance, need for houses, shelter facilities, aid from the government, government visit during and after the flood and their land rights.

Old-Age Allowance

In Bangladesh, the elder people receive an old-age allowance. But these Indigenous people do not get any. Indigenous woman 2 said, "We are old people but none of us get any old age allowance like other people do." According to them, they have no earning source. If they have got any old-age allowance it would help them after the flood. Indigenous woman 1, said, "They have taken our names but said that we are not eligible because of our age. We will die soon. Everyone will see." The government authorities informed us that although our names were taken we were announced ineligible, and this is wrong information. We are totally eligible to receive the allowance, which we do not get.

Building the House

In the flood, the houses of the Indigenous people were damaged, so, they needed to repair the houses immediately. They told us that repairing the house was more important than any other aid. Indigenous man 5 said, "We will need a lot of money to rebuild these houses. We don't have that much resource. It needs money to break it. I think it will cost 30,000 Taka to build the house." His family could not repair their house because they did not have any money. But they know that the government can help them. Indigenous man 6 said, "If the government wishes, they can do a lot like building our house and a shelter house for the indigenous people." Indigenous people could not go to the school as it was used as a shelter because people from the mainstream community already had occupied the place. For this reason, the Indigenous people believe that support from everyone is needed to change the situation. Indigenous man 7 told us with hope, "Collective actions need to be taken. Government should help the poor people especially us, the Indigenous people." As the Indigenous people are more vulnerable, they need additional support from the government.

Shelter

During the flood the Indigenous people needed a shelter, but there were no options for them. Indigenous man 4 said, "Firstly, we need shelter. Then we need food, clothing and home. I took shelter in other people's home." Still, they went to the school where people stayed during the flood, but they got no place for them there. Indigenous woman 5 said, "There were too many people. We did not get any place there to stay. We had to stay on the road because we had no money to rent a room. And a room costs 2500-3000 Taka." The rented room in the higher areas was also not an option because they could not afford them. Some of them went to stay at the school, but they had to return. They were frustrated about their situation and also about the government. Indigenous woman 1 told, "We stayed at the school but water entered into the school too. We have no safety now. There is government but no safety. We could not sleep; we just used to sit all day long." This kind of situation is not good for the physical and psychological health of the people.

Reliefs

During and after the flood, local government provided some relief, but that was not enough. Also, the authority could not cover everyone affected by the flood. The local government did something they should not have done. Indigenous man 4 said, "Nobody came to provide us with any aid. We got some rice and biscuits. Leaders made a list of the flood-affected people, but then erased our names from the list." The people who got something told us what they got during the flood. Indigenous woman 2 said, "We got nothing to rebuild our home. We got some help from them. We got some flattened and puffed rice and a little oil from some individuals. Some people got 4kg rice, oil (250 ml) and other reliefs. But we did not get any." The Indigenous community people got nothing after the flood. They emphasized rebuilding their houses. For doing so the Indigenous man 7 told, "We don't need rations. It's better if they help us to rebuild our home. We need 900 Taka per day to keep a person to build the house. More than 5000 taka is needed to rebuild the house. Where will we get the money?" The Indigenous community people don't have money to cope with the disaster. They have expressed their hope for the government. The Indigenous woman said, "We want the government to help us. Some people got aids, which we did not get. We also wish to get government help." They want a proper and fair distribution of the relief.

Governmental Responsibilities

We asked the Indigenous community people if anyone from the government visited their area. Indigenous man 2 said, "People went to the areas like Mogbazar and Ujanipara. People living in those areas got so many things that they had no place to keep them in their house which we did not get unluckily." They expressed their grievance about this matter. Another reason behind not getting the relief in that particular area is much more political. Ward members only provided relief in their areas only. Indigenous man 6 mentioned, "We live in the 5 No. Ward. We are not supposed to go to the 4 No. Ward and ask for help. If nobody comes to our ward, to whom we will say our sufferings?" It is obvious that inhabitants of 4 No. Ward got the reliefs, whereas the people of the 5 No. Ward got nothing or less than those in 4 No. Ward.

After listening to the stories of the flood of the Indigenous people, we asked what could be done for their betterment. For this question the Indigenous man 5 said, "We don't know anything. We are illiterate people. The government can do everything. The government has the responsibility to take care of us. Government must help us. We need shelter and other essential commodities." It is true that the vulnerable group so much depends on the government, and they are very hopeful. Indigenous man 4 said, "Next time if a flood comes and the Bangladesh government take[s] steps, then our loss can be prevented. There is nothing in our hand. If the Government takes action and helps the Indigenous people only then it can work."

Land Rights

Like other Indigenous communities in Bangladesh, Marma communities have no land rights. Marma Indigenous man 5 said, "Here, life is tough. We have very little means of living. There was a forest, 50–60 years ago. The forest has been cut down. The population increased a lot. Bengalis have settled here. People came from Barisal and occupied the entire place." These people are getting less supply and scope for living in that city where they have been living for more than seven decades. These people do not have any land to live and cultivate. Therefore, they have made their houses near the river because this is a cheaper option. Indigenous man 7 mentioned this, "We have been living here since our birth. We are poor and do not have enough money to pay rent, so we made our home near the river to save some money to educate our children. But flood destroyed our home and took away everything we had." The flood has indeed hampered deeply their life. Marma Indigenous man 2 mentioned, "People are illiterate and ignorant. Indigenous people sold their land here and went to remote areas like Ruma and Thanchi. They have left and it was their own choices." The Indigenous people are not aware of their land rights. So many Marma Indigenous people have sold their land and have gone to the upper hilly areas in Bandarban. Sometimes they had to leave their land due to pressure from the mainstream. The Indigenous leaders are not able to do anything about their rights. Regarding this, the Indigenous man 3 said, "Our land and everything have been taken. Everyone knows that we live in Bandarban but this has no value. Leaders are taking everything in their pockets. Indigenous leaders can do nothing."

Indigenous Land-Based Practices

Indigenous people are really concerned about their identity and their spirituality. The elderly storytellers are well aware of their very existence. Indigenous man 1 said, "We are Buddhist. Everyone has their own religion." They have mentioned their own spirituality and also acknowledged that everyone has their own spirituality. Helping each other, showing gratitude, respecting the elderly, etc., are the common customs of spirituality. Marma community people think that people have gone far from spirituality, that is why trouble is coming toward people and they are suffering. Marma Indigenous man 2 said, "People became sinners, they don't obey religions. That's why floods happen. People are also helpless because people don't respect elderly people nowadays." Actually, if people do harm to nature it will react in an astonishing way. The same thing they have described in their own way. The Indigenous people are also unhappy because of the new generation. They are less interested about the Indigenous practice that will empower them in society. Indigenous man 3 said, "Young generation does not understand these things. They said that what will be the benefits of praying or doing spiritual activities." It is possible that the older people cannot spread their spirituality or Indigenous knowledge among the youth because of poverty and the shrinking situation in the area, as they are surrounded by the mainstream community.

Discussion and Conclusion

From the study, we have understood that 90% of Bandarban City was submerged. Days had passed in most regions where the water had not subsided, and the likelihood of more rain means that the suffering of the local people will not end. The long-term lack of necessities such as food, water, shelter, and security assurances in the far corners of this border zone is enough of a warning of the disastrous events that could occur. To save and rehabilitate individuals who are in danger, the government must thus make a decisive effort. To that goal, as well as to stop any fatalities and lessen the damage, all required administrative staff and resources must be sent out immediately. However, one must also comprehend why the district had been so severely impacted that time around, given that it did not flood during periods of greater rainfall some 20 years ago. That, in large part, had been caused by the environmental devastation of Bangladesh. The soil has deteriorated and softened as a result of the frequent hill cutting and stone removal needed to construct houses. A significant additional factor is the destruction of trees. Flooding has become unavoidable as a result of these circumstances. Meanwhile, river water might overflow readily due to the absence of dredging and the raising of riverbanks. Additionally, since significant areas have taken up residence in several canals that are a component of the district's drainage system, it is difficult for the overflowed water to leave.

In this part, we have tried to find out some issues that the government needs to resolve immediately.

Self-determination

Indigenous community people are more likely to stick with themselves. Through their individual and collective capacities for self-determination, Indigenous people are working toward their well-being in order to realize their common goals of shared prosperity. Indigenous people define self-determination, make use of their individual and collaborative self-determination capacities, and are prohibited from using their self-determination capacities by restrictions on their liberties (Gordon and Datta, 2022). Indigenous people adapt to climatic events by incorporating various components into their society. Some successfully cope with environmental changes, while others struggle. Cultural and religious festivals in CHT communities increase bondage and communal feeling, fostering resilience and providing emergency needs to overcome vulnerability (Garai et al., 2022).

Fear of the Mainstream/Community-Level Vulnerability

We came to know that many Bengalis have been pushed to migrate to the CHTs due to extreme weather events, rising sea levels, soil salinity, and flooding. Due to loss or damage to land, members of many coastal communities, whose livelihoods depend on farming or fisheries, are choosing to move to urban cities or the low-populated hill areas of Bandarban, Rangamati, and Khagrachari (Haque et al., 2021a). These people disturb the eternal ecology of the hills that in the long run will result in the disaster. They are replacing the Indigenous people by grabbing their land, which is not appropriate. And the Indigenous people are losing their customs, culture, and practices, which is really dreadful. Indigenous people are not so comfortable around the settlers; they have some hidden fear for the settlers. This fear also makes them vulnerable.

Government Responsibility

We have learned that Indigenous people usually attempt to find unique solutions to deal with human-created environmental disasters. They create novel concepts to adjust to the harsh circumstances brought on by extreme climate change. Instead of building typical homes, they attempt to create homes made of brick and cement, which helps shield their home from flood waters. Native Americans who lack resources only build their homes' foundations out of brick, sand, and cement. In addition, they construct a two to three-foot wall around their home, with a fence or tin wall built atop the brick wall (Garai et al., 2022). The government needs to include Indigenous populations in the decision-making process. Democratically elected representatives at the administrative level would give people the comfort of knowing their voices are being heard (Rahman & Biswas, 2021).

Indigenous Land-Based Learning

Most of the Bengali settlers who move to CHT have little to no knowledge about the importance of the traditional lifestyle of the Indigenous population on the cultural diversity and environment. Therefore, the government should endorse educational programs for Bengali settlers so they can learn, understand, and appreciate the rich history and traditions of Indigenous peoples and their importance in preserving the natural environment (Rahman & Biswas, 2021). The ethnic community of CHT has traditional knowledge, practices, and technologies, which are unique to their culture and society. The knowledge and practices are environmentally and socially appropriate and sustainable (Ahamed, 2001; Sarkar & Sarkar, 2015) and being transmitted from generation to generation orally (Khan et al., 2007; Mohiuddin et al., 2012). Traditional knowledge supports livelihood and influences lifestyle, land-use planning, resource conservation, and culture unique to a given culture or society. Mohiuddin (2009) documented 26 traditional knowledge practices by the tribal communities where he identified 289 ethnomedicinal and food plants growing in wild habitats in Bandarban. The community has vast experience in the utilization and conservation of biological and ecological diversities and ecosystems (Khisa & Mohiuddin, 2006). Strong ties between humans and the land have resulted in the creation of complex and wide-ranging ecological knowledge, which includes an understanding of the advantages of rain. These skills have been passed down through the centuries (Tahir et al., 2024).

Indigenous people in hill tracts are developing cultural strategies to respond to climate change extremity and environmental change. Governments and policymakers should formulate policies to uplift these people and incorporate modern technology and institutional responses to combat climate-induced hazards. Indigenous knowledge can help stop susceptibility in CHTs and strengthen the resilience and empowerment of impoverished and vulnerable populations. Integrating native early warning systems into locally run warning systems can also help combat climate-induced hazards and improve the lives and livelihoods of these communities.

5 Marma Traditional Land-Based Practice in Building Community Resilience-Perspective from Chittagong Hill Tracts (CHT), Bangladesh

This chapter explores how the Marma Indigenous community in the Chittagong Hill Tracts (CHTs) utilizes land-based knowledge and practices to build climate resilience amid increasing environmental challenges. Unplanned infrastructural development, land grabbing, and deforestation have exacerbated the community's vulnerability to extreme climate events, such as flash floods and water scarcity. The chapter adopts a decolonial approach, centering Marma voices and addressing historical injustices and sociopolitical marginalization. It highlights the community's traditional land and forest management techniques, the impact of anthropogenic environmental degradation, and the diminishing natural resources that threaten their livelihoods and cultural practices. The chapter concludes with community-driven solutions for sustainable land, forest, and water resource management.

Taking a community-led approach, this chapter discusses Marma[1] Indigenous communities' land-based knowledge and practices in building community-level climate resilience. Unplanned infrastructural development, government declaration of reserve forests, and increasing numbers of settlers have made the Marma Indigenous community in CHTs more vulnerable to the intensity of the extreme climate (Iftekhar & Hoque, 2005; Uddin, 2014). The already-marginalized Marma Indigenous community is at high risk for human-induced water-logging, and flash floods are sweeping away their homes and cattle that are severely affecting their livelihoods. Therefore, this chapter focuses on how anthropogenic environmental causes such as stone lifting, land grabbing, rapid deforestation, and waste disposal are causing water scarcity to identify community-led adaptive measures in light of the Marma Indigenous community's land-based practice. The chapter follows a decolonial approach that addresses the limitations of the Western-based research method and centers on Marma Indigenous communities' voice regarding historical injustices, settler's invasion, land dispossession, and sociopolitical marginalization of the communities.

The climate crisis is having a profound impact on the loss of traditional forests and biodiversity, hence the loss of livelihoods of the communities depending on the forests and lands for their livelihoods. The rapid reduction in natural resources is affecting the sustainable traditional cultivation system,

DOI: 10.4324/9781003536345-5

polluting and reducing safe drinking water sources, and causing flash floods and landslides endangering the land-based communities and their existence. Fifty percent of the people of the CHTs, especially the Indigenous communities, are suffering from acute water scarcity (Khan, 2015) due to prolonged flash floods and water pollution that are affecting their lives. The Marma Indigenous community relies heavily on *jhiris*[2] as a source of water, which is now drying up due to excessive stone lifting. With continuous land grabbing by the settlers[3] in the name of administrative and tourism structures, brickfields, tobacco plantations, lumber plantations, and reserve forest projects have diminished the sources of their livelihoods. Furthermore, Marma communities' traditional land and forest management knowledge was diminished by the invasion of government in the CHT region limiting their capacity to respond to climate change (Datta, 2015). In the name of "so-called development," several human-induced activities such as massive deforestation, unplanned land use, and resource management are aggravating environmental degradation and causing the drying up of waterfalls, landslides, flash floods, droughts, and heavy rainfall, among other disasters.

Historically, the Marma people have their own land-based techniques and knowledge better suited to the local ecology and climate. They have developed sustainable land and forest management techniques and passed the knowledge to the generations through their land-based customs and traditions of storytelling, dancing, and singing in their daily life. The age-old traditional knowledge is diminishing due to the government's land policy, illegal invasion, anthropogenic pollution, etc. It is evident that natural sources of mountain water are dying up due to massive deforestation, frequent landslides, and precipitation changes in the CHT areas (*The Business Standard*, 2024). It is also reducing the water flow in the river and affecting the lives of the Marma community dependent on the river for their livelihood. This severe water crisis is pushing the community villagers to migrate to cities. On the contrary, the riverbeds are silted due to deforestation, landslides, waste disposal, and water pollution causing prolonged flash floods that are sweeping away houses and livestock of the Marma community.

There are not enough studies highlighting the sociopolitical subjugation of the Marma community that is destroying their traditional forests and water resources by the colonial forms of governance, thus, destroying their livelihoods and their age-old forests and water resource management practices. The rapid reduction in natural resources is destroying their capacity to adapt to climate and to build community resilience. The government of the CHT is overlooking the natural resource depletion and systematic extinction of the Indigenous communities whose livelihoods are depending on the lands (Roy, 2000a). The mainstream research failed to capture their land-based knowledge and (to a great extent) contributed to the subjugation of the Indigenous communities (Banerjee, 2000). The findings of this chapter focus on community-driven solutions for sustainable land, forests, and water resources management as a tool for community adaptation.

Theoretical Framework and Methodology

This chapter follows the Indigenous land-based research framework from a decolonial perspective that acknowledges the inherent bias within existing Western-based research methodologies and attempts to reframe Western perspectives, fostering a more equitable and respectful engagement with Indigenous knowledge systems (Smith & Sharp, 2012). Aligning with land-based ethics, the decolonial approach emphasizes the importance of centering land-based values, knowledge, and experiences within the research process (Datta, 2018b; Wheeler, 2001). We followed a participatory action research (PAR) method by following a respectful and meaningful engagement with the Marma community in exchanging knowledge and learnings. We used deep listening and storytelling to capture Marma's land-based learning. Traditional land-based storytelling is a significant way of expressing Indigenous knowledge, culture, and oral traditions (Kovach, 2009). Studies suggest that storytelling is an important component of transforming knowledge over generations through folktales, songs, rituals, chants, and even artifacts (Denzin et al., 2008; Kovach, 2009). Following the PAR guideline, we situated ourselves within the participants' community, empowering participants, recognizing the participants' spiritual and relational knowledge, and taking political stands for the participants' community. We used deep listening to exchange learnings regarding climate change impacts and adaptation techniques during our visit to the Marma community in August 2023. Between 20 and 25 Marma community participants, elders, and knowledge keepers participated in deep uninterrupted conversations to share the traditional knowledge that helped them address water scarcity and how settlers' invasion is endangering the ecological balance of the region. Throughout this process, we consistently sought and obtained consent from the community, ensuring a collaborative and ethical approach to developing this manuscript.

This chapter follows three broader themes: the meaning of climate change to the Marma community, human-induced causes of water scarcity and its consequences on the community, and the Marma community's land-based practice as a solution to adapt to climate change. Table 5.1 illustrates the themes and subthemes of this chapter.

Historical Context

The biodiversity of the CHT region has been degraded drastically over the past couple of years (Thapa & Rasul, 2006) due to illegal encroachment and declaring reserve forests. Historically, the Marma community has developed a deep connection with forests that helped them to harvest and to manage forests sustainably. In 1860, the British government interfered in the CHT region to harvest forests and evicted Indigenous communities from their traditional forest (Schendel et al., 2001). The illegal invasion continued until the 20th century in the name of the expansion of reserve forests (Thapa & Rasul, 2006). Approximately 350,000 hectares of land were snatched illegally from the

(a)

(b)

Figure 5.1 How the houses of the Marma community were destroyed due to the flood disaster in August 2023.

Table 5.1 The themes and subthemes of this chapter

Themes	*Sub-themes*	*Examples*
Meaning of climate change	Community perception of climate change	"There is only one river in our locality 'Shankha River' to pass the rainwater. The rainwater cannot pass through the river as the river is grabbed by tourism projects."
	Acute water crisis	"Earlier it took only 20 minutes' walk to collect water from the neighborhood waterfall, but nowadays it takes one hour to climb uphill for water and then another half an hour to climb down from the hill with water."
Human-induced causes of water scarcity	Declaring reserve forests Rapid deforestation Unplanned development infrastructure obstructing water flows	Approximately 350,000 hectares of land were snatched illegally from the Indigenous community."The use and disposal of plastic is increasing drastically due to increasing numbers of settlers every year. It is restricting rainwater from passing down to the river and causing a prolonged flood." "We have no access to our forests and lands; from where do we collect foods? We have to work on the Rubber and Teak plantation projects as labor with a minimum wage. No one is thinking for us."
Consequences of water scarcity	Scarcity of food and losses of livelihood Severe water scarcity Prolonged flash floods/landslides Losses of biodiversity and loss of Marma land-based knowledge	"The extreme heat and prolonged drought negatively impacted last season's crop yield. It caused a shortage of food. In 2023, we faced periods of starvation as a result of the flash flood." "Earlier there was no garbage in *Jhiri* water, but now the water is polluted due to plastic. Community people are suffering from drinking this polluted water."
Land-based practice: a tool to address climate change	Mapping of Indigenous plants Preserving land-based knowledge Integrating Marma land-based education in the existing curriculum	"Our grandchildren do not know the traditional plants and trees. We are the last generation to know about the benefits and the use of plants. If not preserved, our knowledge will die." The Marma community have their land-based knowledge to select crops for high altitude and wind velocity. Marma youth education should reflect a spiritual connection with land, water and traditional sustainability.

Indigenous community ignoring their rights, dignity, and connections to lands and forests (Roy, 2002). The illegal encroachment in the CHT region was at its peak during the military regime in 1989 when the government allowed the Bengali settlers to take control over forests, lands, and rivers. In 1998, the CHT forest division was established to take control of the forests and to restrict the Indigenous community from their sustainable *jhum* cultivation (Hossain, 1998). This approach fails to recognize the role of the Indigenous community, their relationship with lands, and their age-old sustainable land management techniques (Ahammad, 2019).

During the mid-20th century, the monoculture (*rubber, teak, Gamar, Chapalish, Jarul, etc.* in Bangali) was initiated by the Forest department of the CHT region as part of maximizing timber production and profit making (Thapa & Rasul, 2006). The revenue collected from forest harvesting decisions has influenced the massive deforestation of traditional forests and converted them into monoculture (Thapa & Rasul, 2006). The cultivation area in the CHT is currently dominated by teak plantations, occupying nearly 70% of the land, posing a significant threat to the local ecology (Feroz et al., 2014). This attention on cash crops has led to the destruction of traditional forests and the spiritual significance associated with them. The soil fertility and moisture content are decreasing due to monoculture and are causing a loss of biodiversity (Mahmud et al., 2017). From 2000 to 2015, the CHT region witnessed 8% of its forest reduction (Sarkar & Mukul, 2024). As the Marma community heavily relied on the forests for their livelihoods, the destruction of traditional forests is forcing them to migrate to upper hills or cities where they are forced to lead a measurable life with bare minimum facilities. The unplanned growth of tourism infrastructure, roads, and the arrival of Bengali settlers in the CHT region have led to rapid deforestation and illegal land encroachment (Iftekhar & Hoque, 2005; Uddin, 2014).

Findings

The neo-colonial approach taken by the government of Bangladesh in the form of land appropriation, reserve forests, and unplanned infrastructural development has destroyed the ecological balance and exacerbated the vulnerability of the Marma Indigenous communities living in the CHT areas. Climate hazards such as flash floods, droughts, stone grabbing leading to water scarcity, prolonged rainfall triggering landslides, drying up of waterbodies due to lack of rain, and soil erosion are some of the examples of climate-related challenges faced by the Marma Indigenous community. These hazards have severe impacts on the lives and livelihoods of the local Indigenous community as they have less access to finance, education, and basic facilities that are making them even more vulnerable to climate change and limiting their access to their land-based practices. Over the centuries, the Marma community have developed and implemented their land-based practice and knowledge to build a sustainable community centering their culture, practice, and spiritual beliefs. Their land-based practices, deeply rooted in forests, water, and lands, helped them to harvest natural resources in a sustainable way without destroying ecological balance and to adapt to climate change. The settlers' expansion and the government's land appropriation in the name of the development project are destroying natural resources and ecological balance, thus, diminishing the Marma land-based traditional knowledge and practice. The Marma community identified deforestation, land grabbing, stone lifting, and settlers' expansion in the CHT region as the potential causes of environmental degradation that are causing food and water scarcity, landslides, flash floods, forced migration, etc. They have also emphasized youth education should reflect a spiritual connection with land, water, and traditional sustainability. It is high time the government recognizes the Marma land-based practices in sustainable community development and natural resource management. Marma traditional plant mapping, traditional reforestation, integration of land-based knowledge in the mainstream curriculum, and investing in land-based knowledge and practice documentation are some of the proposed measures suggested by the community to preserve land-based knowledge.

Community's Perspectives on Disaster

During deep listening with elders and knowledge-keepers, we observed that the Marma community is affected by the destructive outcomes of human-driven development activities that are diminishing their lives and livelihoods. A Marma elder 1 identified the colonial forest management policy, government-backed settlers' invasion, and unplanned development activities as the dominant causes of environmental degradation.

The elder 2 Mongchu Marma explained how they are being deprived of accessing their traditional lands due to government-initiated monoculture. He said, "We have no access to our forests and lands; from where do we collect food? We have to work on the Rubber and Teak plantation projects as labor with a minimum wage. No one is thinking for us."

During the storytelling, the Marma elder 4 explained that the use and disposal of plastic is increasing drastically due to increasing numbers of settlers every year. It is restricting rainwater from passing down to the river and causing a prolonged flood. We have learned that the tourism infrastructure is ignoring environmental protection and disregarding the interests of the Marma community. The unplanned construction at the hill slope is contributing to landslides. Also, land and river grabbing and disposal of plastics are causing water logging and leading to prolonged flash floods.

Marma Community's Perception of Climate Change

The Marma community has their own land-based rituals to honor the oldest trees and waterbodies as the protectors of their community. They have a long tradition of worshipping their own deities associated with the land, rivers, forests, crops, animals, and other living and non-living entities, demonstrating a strong connection with the land and forests that shape their daily lives.

During storytelling, we observed that the Marma community is experiencing the devastated consequences of settlers' invasion in the form of acute water scarcity, prolonged flash floods, and related water scarcity. A Marma elder 1 said, "There is only one river in our locality 'Shankha river' to pass the rainwater. The rainwater cannot pass through the river as the river is grabbed by tourism projects." The marginalized people of the Marma community are migrating to cities as the drying up of rivers is reducing the fishing opportunities from the river. During the dry season, the Marma woman has to walk up hills to collect water. Mongchu Marma elder 2 said, "To us, climate change means the powerful people are destroying lands and forests and thus destroying the sources of our livelihoods."

Marma Community: Human-Induced Water Scarcity

Elder 2 Mongchu Marma said that in their childhood the water holes (that contain *jhiri* water) used to be full of water all year, but nowadays even in the rainy season water is scarce. During the community engagement, the Marma community said that they have experienced less precipitation during the past couple of years because of massive deforestation. Marma women said that water scarcity is impacting their health badly. During the dry season, they have to walk miles uphill at midnight to collect water for cooking and household use. We found that the precipitation rate in the hills was reduced from 2,500 mm to 3,000 mm annually to 1,500 mm, which is reducing water flow in the river that is alarming for the livelihoods of the local people. The Marma community woman elder 3 said, "Earlier it took only 20 minutes' walk to collect water from the neighborhood waterfall, but nowadays it takes one hour to climb uphill for water and then another half an hour to climb down from the hill with water."

They have to start collecting water at midnight to avoid extreme heat in the daylight. The Marma community has identified settler explosion in the hills as the dominant cause of water depletion and water scarcity. They have said that the acute water scarcity is skyrocketing the price of water. Earlier they had to pay BDT 600 per 1,000 liters of water, but nowadays it has increased to BDT 800 per 1,000 liters of water. The Marma community also believed that the use of pesticides and chemicals in the cropland is polluting the waterfalls and rivers and is killing the fish and other species that live in the water. Since the Marma community's life is dependent on water, water pollution is reducing the means of their life and livelihood.

The ongoing water crisis was aggravated during August 2023 when the Marma community experienced a devastated flash flood. During deep listening, the Marma elder said that their houses and cattle had been washed away during the flood, and they had to live on the high roads. They had to drink rainwater that caused skin infections, stomach diseases, sanitation and hygiene crises, etc. The community people lost their crops and sources of income during the prolonged flash flood. The Marma community elders believed that deforestation, river grabbing, and waste dumping were hindering the regular flow of rainwater and causing prolonged flash floods.

Consequences of Human-Induced Waterlogging and Environmental Degradation

This section will discuss the impact of anthropogenic activities, illegal encroachment, and massive deforestation on the livelihood of the Marma community and the loss of their land-based practices that diminish their capacity to respond to climate change. During the storytelling, the Marma community shared the consequences of environmental degradation on the Marma community's lives, livelihoods, and ecological balance.

Scarcity of Food and Losses of Livelihood

During the deep listening we have found that the Marma community is currently dealing with crop damage due to extreme weather events like drought, declining groundwater levels, and shifts in precipitation patterns. The community is being impacted by intense heat and heavy rainfall, which are disrupting their agricultural practices. Additionally, prolonged flash floods are leading to the loss of young plants. In August 2023, the Marma community suffered crop losses due to a flash flood. Due to prolonged floods, they lost the seedling planting season of rice. Traditionally, the Marma community were rich in agricultural production. They are using their land-based knowledge from their forefathers in sustainably cultivating crops. The ongoing anthropogenic activities in the name of so-called "development" are reducing the Marma community's access to lands and forests and leading to food and livelihood scarcity in the community. Mongchu Marma elder 2 said, "The extreme heat and prolonged drought negatively impacted last season's crop yield. It caused a shortage of food. In 2023, we faced periods of starvation as a result of the flash flood." The elder 5 said that their homes and cattle were washed away during the flood. The Bengali settlers were prioritized in the cyclone shelter. They did not have access to go to the shelter. During the night when the water level rose up, they had to stay under the open sky.

Severe Water Scarcity

The Marma community depends on *jhiri* for their sources of water. They use the *jhiri* water for drinking, bathing, cooking, and household activities. The illegal stone lifting from rivers is creating a scarcity of water resources. Furthermore, various anthropogenic activities such as garbage dumping and land and river grabbing are polluting water sources and disturbing the regular water flow. Chaw U Prue Marma said, "Earlier there was no garbage in

Jhiri water, but now the water is polluted due to plastic. Community people are suffering from drinking this polluted water."

During the participatory discussion, the elder Marma said that rivers are becoming muddy all the time, and there is no source of freshwater. The Marma community people said that the stone lifting and illegal constructions are run by the local political leaders to expand their business profit ignoring the interests of the community. The Marma community faced a severe water crisis during the recent flash flood in 2023 that has aggravated the health and hygiene condition of the Marma community. The elder Marma said that they had to drink rainwater during the flood because their traditional sources of water became polluted. Every rainy season the Marma people suffer water-borne diseases.

Prolonged Flash Flood

The Marma communities saw arguably the biggest flash flood ever in 2023. River pollution, frequent landslides due to rapid deforestation, dam construction, and land grabbing have worsened the situation by blocking the rainwater flow, thus, causing prolonged flash floods that have affected the Marma community severely. Need Assessment Working Group hosted by CARE Bangladesh in its assessment report estimated that 1.2 million people and 34,973.5 hectares of crop land in the CHTs were affected due to the flash flood. The Marma community that lives in remote areas of Bandarban District is even more vulnerable to disasters. All communications were cut off during the flood. They had to starve for weeks. Their homes and cattle were washed away. They had to drink rainwater that caused stomach diseases. The Marma day laborers could not get any work, and hence could not buy food, medicine, etc. for their families. They were being discriminated against to receive relief services. Marma Elder Mongchu said, "The political leaders are giving relief to the settlers. We have received only 5kg rice which is a mockery in the name of giving relief."

The long-time ignorance from the state that directly or indirectly prompts the environmental degradation in the form of unplanned development works, land grabbing and exercising power over the Indigenous communities, and massive deforestation is causing water pollution and waterlogging.

Landslides

In recent times, landslides have become very frequent in Marma communities and are becoming life-threatening for them. An elder Marma said that the landslides are becoming a frequent phenomenon due to the rapid expansion of settlers in unplanned ways and ignoring the land rights, and traditional housing techniques of the Marma community. Monoculture and brickfields are also contributing to frequent landslides. During the storytelling, many of the Marma people said that they had escaped from landslides. Mongchu Marma said, "The hilltop soil is becoming loose due to landslides and deforestation. Nowadays we are very afraid of landslides."

Loss of Biodiversity and Loss of Marma Land-Based Knowledge

The expansion of settlers in the CHT region poses a major threat to the ecological balance. The elder Marma believed that the forests and lands being degraded and exploited are causing a huge loss of biodiversity, local plants, and species. The Marma community, directly and indirectly dependent on forests for their livelihoods, is becoming more vulnerable. The heatwaves, changing temperatures, and precipitation have severe consequences for species that are climate-sensitive. 20% to 30% of the species are at the threat of extinction due to unsustainable exploitation of natural resources (Pender, 2008, DoE, 2015). Marma community's life, cultivation, and spiritual practices revolve around forests and lands. The land-based practice, a proven tool to address community resilience, will diminish gradually due to the loss of forests and their related biodiversity. The elder 6 said, "Earlier we used to make our own medicine from leaves, and branches of Indigenous plants that are now being extinct due to settlers' invasion." The elder Marma emphasized educating youth on the Marma land-based practice to preserve and promote Marma land-based practice.

Environmental degradation has a long-lasting impact on the lives and livelihoods of the Marma community including food and shelter crises, water scarcity, flash floods, landslides, losses of crops, etc. However, the ignorance of the government has aggravated anthropogenic activity-induced environmental degradation and has worsened the situation by increasing the severity and frequency of disasters such as flash floods, waterlogging, and landslides.

Marma Land-Based Knowledge: A Tool to Build Disaster Resilient Community

The land-based techniques and community bonding are proven to adapt to climate change. For example, the house-building techniques on the slope of the hill, the shifting cultivation process, and traditional techniques to collect and preserve water are very helpful in responding to disasters. The food and seed storage techniques of the Marma are also a critical aspect of disaster management. The oppression from government discrimative policies, illegal power exercise, and Bengali settlers' expansion in the CHT region are contributing to unsustainable harvesting of natural resources and are leading toward the loss of biodiversity.

The rich biodiversity of this region is the home of Indigenous communities for their dependency on lands and forests. The Marma community have developed strong bonds with lands through their daily and spiritual practices and rituals without destroying the ecological balance (Adnan, 2004; Mohsin, 2002; Roy, 2000a). The cultural identity, the storytelling, dancing, singing, and

Figure 5.2 shows how the houses of the Marma community were destroyed due to the Flood Disaster in August 2023.

worshipping of lands and forests used to be the vehicle to pass down the sustainable land management practice to the next generation where the elder acted as the knowledge keeper and the youth acted as the learner (Datta, 2015). However, the government is failing to recognize the importance of Indigenous land-based practice in sustainable lands and forest management. The formal education system is excluding land-based knowledge from its curriculum and hence is failing to design techniques sustainable for community disaster management. This exclusion of land-based education is alienating the Marma youth from their traditional knowledge and practices.

The Marma community proposed some adaptive measures for community-level disaster preparedness and recovery in light of their age-old land-based practices. The Marma community emphasized the government and policy adaptation to preserve and promote their knowledge to build community resilience. The proposed measures include the following:

Mapping of Indigenous Plants

The elder Tapan Kanti Marma emphasized the documentation of Indigenous plants and species and their benefits in providing livelihood and ecological balances. The Tapan Marma elder 7 said, “Our grandchildren do not know the traditional plants and trees. We are the last generation to know about the benefits and the use of plants. If not preserved, our knowledge will die.” It is reported that 225 plant species were extinct in Bangladesh during the past 100 years mainly due to anthropogenic causes. Local variations in crops are also declining due to the promotion of high-yielding profit-making crops. Local plants have a significant contribution to food security, and they are well-adapted to the local climate (Sarwar, 2019).

The Indigenous community is highly dependent on forests and lands for their livelihoods. Thus, the continuous loss of biodiversity is weakening the connection of the Marma community with lands and forests and thus aggravating natural resource depletion. Considering the situation, many countries have adopted Indigenous plantations on the slope of the hills and riverbanks as a technique to integrate flood management (Integrating native planting and flood protection, 2021). It is urged by the Marma community that the government should prioritize the mapping of traditional plants and species and promote the plantation of these traditional plants.

Preserving Land-Based Knowledge

The Marma community has land-based rituals to honor the oldest trees and waterbodies as the protectors of their community. They have their deities related to lands, rivers, forests, crops, animals, and other living and non-living entities that shape their daily lives (Chakma, 2010b). According to elder 8, most of the worship is performed as a collective effort to make a stronger community bonding. The worship is a celebration of land-based knowledge where the elders pass down the knowledge to the young generation.

This research shows that the Indigenous community's traditional land-based knowledge is effective for community resilience. The Marma community has innovative ways of farming to adapt to climate change. The Indigenous communities in CHT are cultivating 55 crop species in *jhum* techniques that are sustainable for ecological balance (Datta, 2015). The local crop variants are tolerant to cold and hot weather (Mohiuddin, 2009). The Marma community have their land-based knowledge to select crops for high altitude and wind velocity. However, the cashew nut (cash crop) is not a suitable crop to grow in the high wind. The Marma community have their own water preservation techniques as well. Thus, preserving land-based knowledge will also help the government departments to adopt sustainable solutions.

Integrating Marma Land-Based Education in the Existing Curriculum

It is evident from the storytelling that the invasion of government by the Western education system is contributing to the elimination of Marma land-based knowledge. The Marma youth is losing connection with the community elders and with their traditional land and nature-based practices and losing ties with their community. The Marma community believes that their land-based education should be promoted to the youths to equip them with traditional cultivation knowledge, Indigenous plant identification, and land and water-related spiritual practices. Youth education should reflect a spiritual connection with land, water, and traditional sustainability. That is why it is the need of the time that the education curriculum should reflect traditional land-based practices. The Marma urged that their land-based practice should be at the center while designing an integrated flood management system for the Marma community.

Proposed Indigenous Land-Based Mitigative Solutions

The traditional forests and lands are interlinked with the life and livelihoods of the Marma Indigenous community to the extent that it is almost impossible to separate them. As the keeper of the lands, the Marma community developed their land-based knowledge for sustainable resource management and community development. The community interactions played a crucial role in transforming knowledge from elders to youths through storytelling, singing, worshipping, spiritual practices, etc. Thus, it is the need of the hour to protect and promote Marma land-based knowledge and practice to develop community resilience against disasters such as flash floods, landslides, and droughts. This section will propose some mitigative solutions to revive the land-based practice of the Marma community to build community resilience against human-induced disasters.

Land Rights

During our deep listening, the Marma community identified that land ownership should be transferred to the Indigenous community for the better management of lands and forests. It is evident that the government's discriminative initiatives are promoting unsustainability in the Indigenous communities (Datta, 2015). The Marma community identified that settler expansion, illegal power exercise, rapid deforestation, and water resource pollution are the biggest causes of natural resource depletion. The existing policy legislation under the Village Common Forests (VCF) program does not allow community ownership of lands and forests. That is why it is high time government initiatives should be taken to redistribute lands rights to the Marma community, which will ensure representation in decision-making, economic empowerment, and access to basic facilities.

Recognition of the Marma Land-Based Practice

Acknowledgment of the Marma land-based knowledge is very crucial for developing sustainable land management practices. Successful integration of land-based knowledge largely relies on the willingness of the government. More research, dialogue, and talks should be organized to acknowledge and promote the importance of land-based knowledge. Tapan Kanti Marma suggested that there should be a dedicated day to celebrate the land-based knowledge of the Indigenous communities and the importance of preserving the land-based knowledge to address climate change. Marma community's life revolves around the traditional lands and forests which is why recognition and promotion of the Marma land-based knowledge will help practice sustainable land management.

Reforming Educational Curriculum

The Marma youth are being digressed from their age-old land-based knowledge and practices due to the invasion of the Western-based singular curriculum. The existing homogenous education system is promoting Western ideologies by ignoring the contribution of land-based practice in preserving lands, waterbodies, and forests (Barua, 2004). The Marma traditional knowledge is diminishing due to the gradual death of their language. Representation of the Marma people as teachers, advisors, researchers, and community engagement leaders is required to underscore the importance of their land-based knowledge. It is high time the government recognizes the importance of incorporating land-based education in the mainstream curriculum to educate the future generation regarding sustainable land management.

Documentation of Traditional Land-Based Knowledge

An elder Marma said with deep regret that they do not have any local or national-level database to preserve their land-based practices. The traditional land-based knowledge is efficient with innovative ideas to tackle climate change. That is why the government should prioritize land-based databases that will help policymakers design and implement the right policies and community-based adaptation techniques to preserve lands and forests. Unfortunately, the funding is very limited for the Marma community's land-based research. The government and nongovernment research organizations should allocate funds to promote land-based knowledge and practices.

Conclusion

This research documents that effective implementation of land-based knowledge is crucial for making evidence-based decisions in building a climate-resilient community. The Marma land-based practice is a way of being and a way of connecting with land that has a deep connection to nature and the ecosystem. By implementing traditional land-based techniques every day, the Marma land-based practices prioritize the welfare of community development through sustainable resource management. For instance, the Marma community experienced the most devastating flash flood in August 2023. This flood resulted in significant losses in terms of lives lost, crops lost, water scarcity, health, sanitation, and hygiene issues. During their narrative, the Marma elders stated that rapid deforestation, environmental degradation, and monoculture are causing natural resource depletion and in ways diminishing Marma land-based knowledge and practices. Given the context, the Marma elders proposed that traditional plant mapping, documentation of land-based cultivation and water preservation techniques, and integration of land-based knowledge in the mainstream educational curriculum are the effective ways to preserve and implement their land-based knowledge. This research suggests that redistributing land rights, recognizing and documenting Marma land-based practice, and reforming educational curricula are some of the mitigating measures to address this issue.

Notes

1 The population and housing census of 2022 indicates that Marma is the second-largest Indigenous community in the CHT, with 224,261 residents. It is believed that in the 14h century, the Marma community's ancestors traveled from the Arakan

State to the CHT to escape the persecution they were subjected to by the Burma (Myanmar) government (Barua & Ando, 2002: 67–68). The rich biodiversity of these CHT forests, hills, and plains used to be the sources of food, revenue, and a means of subsistence for the Indigenous community (Roy, 2000a).

2 A creek narrow stream.

3 The Bengali settler's invasion was popularized in the CHT area during the military regime in 1989. The then government allowed the Bengali settlers to take control over forests, lands, and rivers (Hossain, 1998).

6 Climate Change and Women's Perspectives in the Coastal Areas in Bangladesh

This chapter investigates the impact of climate change on Munda Indigenous women in coastal Bangladesh, highlighting their unique vulnerabilities to disasters like cyclones, floods, and saline intrusion. The study emphasizes the socioeconomic challenges and gendered disparities faced by these women, who are often left behind when men migrate for work. It also explores their community-led coping and adaptation strategies. Centering Indigenous voices, particularly those of women, the chapter aims to inform policymakers about the need to reconsider the fundamental human rights of the Munda people and develop more inclusive climate resilience practices and policies.

The Bay of Bengal experiences almost five times as many cyclones as the Arabian Sea, which increases the vulnerability of India's east coast. The north Indian Ocean basin is one of the least violent cyclone/hurricane basins in the world (Rahman et al., 2023; Sıddıq et al., 2018; Roy, 1998). Bangladeshis, especially Indigenous and minority people who are more susceptible to climate disasters, are the country's most vulnerable inhabitants to tropical cyclones, with an average of four occurring annually, according to the United Nations (Islam et al., 2022; Taru, 2005). Bangladesh's coastal Indigenous communities are particularly vulnerable to natural calamities like storm surges and cyclones. Moreover, additional regular hazards that hurt the fishing and agricultural sectors and cause large losses in terms of lives and property include erosion, floods, droughts, and severe tidal inundation (Datta & Kairy, 2024). The Munda people, who are recognized as Indigenous tribes, are vital to the country's social and economic advancement, which depends on the cooperation of all socioeconomic levels (Sharmeen, 2013; Ahmed et al., 2016). They struggle against several human-created environmental disasters, including heatwaves, cyclones, floods, saline incursion, famines, and the COVID-19 epidemic (Roy, 1998).

Due to societal prejudice, the Munda population in Bangladesh's southwest experiences severe deprivation and neglect; this erodes their culture and deteriorates their language, resulting in superstitions. Despite this, Munda's presence is denied by the government, the majority of Bangladeshis, and the international community (Akinola et al., 2019; Rahman et al., 2023). Women of the Indigenous community suffer a lot from being vulnerable, and they are more vulnerable to dangerous conditions because of their roles as primary resource

DOI: 10.4324/9781003536345-6

users and their dependence on natural resources for their livelihoods (Enarson & Chakrabarti, 2009; (Sharmin & Islam, 2013). Different vulnerabilities cannot be fully explained by gender alone. In local society, disparities in wealth, religion, and ethnicity all support an unequal power structure. To support their families, men are increasingly migrating—or at least motivated to migrate—to the capital city or other neighboring urban centers, according to mounting data from coastal Bangladesh. Women frequently encounter human-created environmental disasters like tropical cyclones compared to men who have moved inland since they are left behind in their native towns in coastal regions (Karlsson, 2020; Chen & Mueller, 2018). The recurring situations that lead to gendered differences in how the climate crisis is experienced in coastal Bangladesh not only draw attention to the social and environmental justice aspects of climate change but also shed light on who, and why, needs the greatest assistance during times of crisis (Ahmed & Eklund, 2021).

In this study the women of the Munda Indigenous community's experience will be shared to encourage the policymakers to reconsider the Munda people's fundamental human rights and contribute them the action they deserve. Furthermore, the Indigenous women provided all of the recommendations for this chapter, which is the voice of Indigenous people. This chapter seeks to accomplish two main goals: (1) find out the climate disaster and various effects of climate change on the women of the Munda Indigenous community living in coastal Bangladesh and (2) explore the coping and adaptation mechanisms led by the community.

Our goal is to comprehend the effects of climate change and the adaptation tactics used by Bangladesh's Indigenous communities in coastal areas, with a particular focus on the extreme vulnerability faced by Munda Indigenous women. This study is a component of a larger study aimed at developing practices and policies for Bangladesh's minority and Indigenous populations. We applied the knowledge that was acquired from the stories of the Indigenous women, who discussed their views on concepts related to climate change, the difficulties posed by weather-related incidents, and their adaptation strategies (Akinola et al., 2019). The study begins by outlining our positionality and then goes on to discuss the theoretical framework that directs our research goals.

Before providing the findings of our data analysis, which we have divided into primary and sub-themes, we describe the study's theoretical framework and methodology. We conclude with comments on community-based adaptations and tactics based on our lessons learned, drawing from the challenges and recommendations expressed by the Indigenous women storytellers and our study's findings.

Researcher Positionality

The role of researcher positionality is of utmost importance in Indigenous research, as it significantly impacts the research process and raises ethical considerations (Datta, 2018b; Wilson, 2008). It is essential to recognize that

Indigenous research frequently delves into sensitive subjects and engages with marginalized communities. Researcher positionality encompasses their cultural background, personal experiences, and social identity, all of which can profoundly influence their interactions with Indigenous participants and their comprehension of the issues at hand. Acknowledging and understanding one's positionality is crucial for approaching research with humility and respect, reducing the risk of perpetuating harmful stereotypes or colonial attitudes inadvertently. It empowers researchers to establish trust within Indigenous communities, nurture meaningful collaborations, and formulate research questions that are culturally sensitive and aligned with Indigenous worldviews.

In addition, our positionality helps to address historical power disparities in Indigenous research. Instead of being active participants or collaborators, Indigenous communities have frequently been the focus of research. We were able to redefine our research agenda by making sure that the research serves the needs of the community by critically analyzing our positionality. Establishing ethical, respectful, and empowering research relationships with Indigenous communities and advancing knowledge that truly serves their interests and well-being are contingent upon our positionality.

Theoretical Framework and Methodology

In the context of disaster research within Bangladeshi wetlands, decolonial feminist research methodology plays a crucial and profound role (Shefer & Bozalek, 2022; Wijsman, & Feagan, 2019; Väyrynen et al., 2021). Its historical past and rich cultural diversity have been disproportionately affected by calamities, which frequently worsen preexisting inequality (Väyrynen et al., 2021). Decolonial feminist study recognizes that women, especially those from marginalized areas, carry a heavy burden during catastrophes and that gender, power, and culture are intertwined in the discourse around disasters (Chapola, 2022). Through emphasizing the perspectives and experiences of women in disaster-affected communities, this methodology aims to reveal resilience tactics and gendered vulnerabilities that are frequently hidden in conventional research (Chapola, 2022). It also emphasizes how crucial it is to honor and strengthen local communities during the process, especially those who have traditionally been excluded and silenced. Decolonial feminist study offers a thorough framework that questions colonial legacies and power structures, which makes it a useful method for comprehending and resolving disasters in Bangladesh's wetland areas.

Moreover, the decolonial feminist perspective is crucial for studying disasters in Bangladesh's wetlands because it acknowledges the intricate network of systemic inequalities and past oppressions that mold the disaster terrain (Datta, 2024a; Wijsman & Feagan, 2019). This methodology examines not just the immediate effects of disasters but also their underlying causes, which are frequently entrenched in social, economic, and political hierarchies, going beyond typical research paradigms (Datta, 2024a).

It highlights the need for community-driven, context-specific tactics that take into consideration the intersectionality of identities and experiences and highlights the shortcomings of top-down, technocratic solutions. Through discussion with local people, especially Indigenous and minority groups, decolonial feminist research addresses their specific vulnerabilities and needs, promoting a more ethical, egalitarian, and inclusive approach to disaster management. The ultimate goals of this methodology are to correct errors in history, challenge hegemonic narratives, and foster resilience—not only in the face of calamities but also in the continuous fight for justice and equity in Bangladesh's wetland areas.

Following the decolonial feminist research framework, we used deep listening and traditional storytelling. We learned stories from nine Indigenous women in August 2023 while abiding by conventional Indigenous customs and ceremonies. Deep listening was an essential strategy for respecting and honoring the opinions of Indigenous women. An important means of expressing Indigenous knowledge, culture, and oral traditions is through traditional story telling. More significantly, traditional storytelling has a good effect on practice and values, spirituality, humility, reciprocity, teamwork, and holistic interconnectivity (Kovach, 2009). The use of traditional storytelling promotes understanding, acceptance of the strength found in both our Western and Indigenous world perspectives, cooperation, inclusiveness, trust, and respect (Datta, 2017). In addition to using deep listening and traditional storytelling, we used reflective writing to think about our research findings and experiences more extensively. We maintain these works to allow ourselves autonomy and the freedom to think freely.

Finally, we can use our thoughts and critical rethinking to sort and value our findings. This procedure enhances our concentration and improves the quality of our research. We strongly tried to maintain the ethical protocols. We maintained the privacy and confidentiality of the respondents and showed our respect and gratitude to them. We kept the privacy of the individuals because they are members of a vulnerable group, the Munda Indigenous community. We also made sure that the participation of the respondents was well informed and voluntary.

Findings: Learning Reflections from Munda Indigenous Women

The majority of the Munda Indigenous communities reside in areas of mangrove forests in the Sundarbans. Originating in India, the Munda Indigenous group came to Bangladesh 300 years ago and has stayed there ever since. They are found in Khulna, Jessore, and Joypurhat, and mostly in the Shyamnagar upazila of Satkhira District, close to the Sundarbans. They are inhabitants of the Sundarbans mangrove forest, and their means of subsistence are derived from the environment. We (authors) as a research team visited a few Munda villages under the unions of Vetkhali, Dhatinakhali, and Burigoalini while conducting this study in the Shyamnagar upazilla.

The researchers listened to the recorded stories several times, concentrating on different aspects of them. The researchers concentrated on the major issues and subtopics that we pointed out. After listening to the recordings a few times, the researchers took some time to reflect on what we had learned.

Compared to the written transcripts, listening to the recorded interviews offers a deeper understanding of a subject. The emotions of the listener are addressed uniquely and more strongly. Our learning featured emotional and often challenging accounts from Indigenous women about their opinions and experiences. The choice of quotes and anecdotes to utilize is often influenced by personal preferences. The majority of the lines that the researcher chose were memorable, thought-provoking, and important enough for us to consider and compel memory.

Main themes and subthemes were developed as a continual form of listening and reflective learning (Table 6.1). The key themes are, first, the Indigenous women's perspective of climate disaster; and, second the impact of climate change on Munda Indigenous women. Following that, we concentrated on the community-led solutions proposed by the Indigenous women.

Women's Perspective on Climate Disaster

The climate disasters described here were shared by the Munda Indigenous women. They mentioned salinity, food crises, floods, cyclones, drought, and water crises as climate disasters because they have faced them in their area. Now we will discuss their stories of these disasters.

Salinity

In the coastal areas, salinity is a major issue for the inhabitants. Their daily lives get hampered by this. They are also aware of the impact of climate

Table 6.1 Main Themes and Subthemes of the Results

Women's Perspective on Climate Disaster	*Impact of Climate Disaster*	*Community-led Solutions from Indigenous Women*
• Salinity • Food Crisis • Floods • Cyclones • Drought • Water Crisis	• Displacement (Housing, Land) • Health (Sanitation, Existing Treatment Procedure, Personal Hygiene) • Education vs. Early Marriage • Social Stigma • Inequality in Work Place • No communications with Policymakers	• Women as Change Makers • Indigenous Leadership and Women's Participation • Land Rights and Women's Rights • Awareness Buildup Within and out of the community • Language and Cultural Rights

change. Munda Indigenous women-1 said, "The effects of climate change are numerous. We don't get fresh water during high tide; instead, saltwater enters in our area. There are shrimp quarries everywhere you look." When the saline water enters their area, scarcity of drinking water happens. In that area shrimp farms are the main livelihood of the mainstream people. According to the Indigenous women, the farms are also worsening the salinity. Munda Indigenous women-2 said, "There is only one freshwater pond here, and rest of the water is salty. Shrimp farms can be found anywhere. Saline water enters our freshwater pond when they create a water entry point on the farm." The Munda Indigenous women also realized this. Salinity is also hampering crop production and having a negative impact on the health of both men and women. The Indigenous women-3 mentioned the fact, "Our freshwater pond is becoming less fresh and more salted. Our ovaries become infected because we use saline water. We are unable to grow vegetables. Not even rice can be grown here." Therefore, salinity has an impact on crop production, livelihood, health, availability of fresh water, etc.

Food Crisis

Due to the lack of work, they cannot earn money and buy proper meals needed for a human being. For this reason, the Munda Indigenous women-4 said, "We are unable to eat a healthy diet. Vegetables from the market are poisonous if we purchase them. However, we are unable to cultivate veggies here. We would not use poison or fertilizer if we could grow vegetables here." One thing is mentionable. Though they have a food crisis, they are aware of the environment. They have a strong relationship with the land, water, and nature, and they will not do any harm to nature. The food crisis goes at its peak after every disaster. They have to live only on dry food like flattened rice, puffed rice, jaggery, etc. When a food crisis happens after any disaster, they become more upset. The Indigenous women-5 said, "A food crisis is here. How many days can you live on dry food and water after every disaster? At the time of the disaster, people start to worry and get unwell easily."

Flood Disaster

Flash flood is another disaster in the coastal areas. Regarding this matter, the Indigenous women-5 said, "Flooding is caused by extreme weather and a lot of rain." They suffered a lot due to floods. The Munda Indigenous women-6 told, "When the flood arrives unexpectedly, we are unable to eat or bathe properly. The roads are underwater. Since we don't have a washroom, we bathe in the neighboring ponds. We either travel or stay at home." The cyclone center is not also approachable by the Indigenous people. They find alternative hiding space during a flood or cyclone. Indigenous women-7 said, "The cyclone center is quite far away. When the flood arrives, we head to the school here. Not even the cyclone center is accessible to us. Though it gets occupied before we get there, everyone lives there." Though the children of the Indigenous children are less likely to get education, the flood also hampers the study of the children." Munda Indigenous women-4 said, "The children are unable to attend school during the flood. During the flood, they stay at home." The floods increase their vulnerability. Every flood makes their life more difficult. The Indigenous women-6 said, "Every time there is a calamity, we run into problems. The cattle suffer greatly when our house is underwater." Not only the humans but also the cattle suffer due to the flood.

Cyclones Disaster

Cyclones are more common in the coastal areas of Bangladesh. They face more than five cyclones in a year. Cyclones also worsen the situation. Munda Indigenous women-6 said, "Our house is broken. Cattle and poultry are washed away. We become hopeless after losing everything. I lost mine during Cyclone Aila I built my new house after the storm." During a disaster, the Indigenous people cannot go to the cyclone center so they stay in their area. The Indigenous women-7 said, "We used to stay on the roads throughout the cyclones. We cannot go anywhere. Sometimes we take shelter at the village's primary school." The Munda Indigenous people suffer a lot during a disaster. Also, they have nowhere to go. Munda Indigenous women-8 mentioned, "So many big typhoons hit us so far, but we never leave our land." The Indigenous people feel that the land they are living on is a blessing to them so they never leave their land behind.

Drought Disaster

The coastal areas are very warm, and the land is not fertile. The extreme weather strikes them very badly. The Indigenous women-8 said, "Heatwaves and drought are also common in here. If we go to work, we become sick." The people here also get sick when they work in the saline water or field or as day labor during the heatwaves. The woman added, "During the summer you have a scarcity of drinking water." The Indigenous people in the coastal areas barely have any permanent work. During the extreme weather, they could not earn money and the food crisis started. Also, various health problems may occur at that time, which increases their suffering.

Water Crisis

In the coastal areas, there is a lack of drinking water. They need to collect drinking water from the nearest freshwater bodies. Munda Indigenous woman 7 said, "This place lacks drinking water. The saltwater is everywhere. The only freshwater body in the area is a pond, from where we collect drinking water." But during a disaster drinking water is very difficult to get. Roads are underwater, so people cannot move. Indigenous woman 3 said, "We filter the water from the freshwater pond to get drinking water, but during the flood, we need to travel great distances to get drinking water. We use boats during the water log." Different disasters bring different sufferings. But their lives go on. An Indigenous woman expressed her feelings, "We experience floods, cyclones, high tides, soil and river erosion, and many more here. People are affected by these calamities but we are helpless too."

Impact of Climate Disaster

Displacement from Housing and Land

Indigenous people in the coastal areas get displaced from their habitat due to various disasters. And once they are displaced, they never return to their land. They lose their house and land. Indigenous woman 1 said, "Thirteen years ago, we moved away from our neighborhood due to cyclone Aila. People who have land make their homes on their property after the cyclone. The other landless individuals left the area like us." They live on other people's land, and they know that they have nothing. The Indigenous woman said, "We don't own any land." This sounds very simple, but this is not so easy for them. They become sad seeing the house of the people of the mainstream community. Indigenous woman 5 said, "People own land everywhere. However, we live on other people's land; we don't have our own land. We are unable to respond to the question of people when they ask us whether we own any land or not." Housing is a person's basic need. They cannot fulfill this very simple need. This is inhuman. But they are bound to live like this forever. The system over there has made them live like this.

Health Impacts on Sanitation, Existing Treatment Procedure, Personal Hygiene

Health and proper treatment in the coastal areas of Bangladesh is no less than a disaster. The Indigenous people are in real trouble if someone gets sick in the community. There are so many reasons behind this. First of all, access a good health center or nearby town is not easy. The Indigenous woman 4 said, "We cannot travel to the hospital when someone becomes ill, especially if they are pregnant. Safe delivery is not ensured here; we must travel to Shyamnagar for it. We are literally in a crisis of life and death." Also, although the treatment is cheap or free, they are bound to pay a very high price for transportation. Regarding this matter, the Indigenous woman 8 said, "We visit a clinic close to the Upazilla Parishad. The patient receives free consultations from the doctor and free medication. But we have to pay more for the transportation because it is so far away from here."

The area the Indigenous people are living in is not appropriate for health. Every now and then they suffer from various diseases. But the women suffer more than their male counterparts. Salinity, poor nutrition, and lack of fresh water make the situation worse. Indigenous woman 7 said, "Salt causes a number of problems for women. Women experience irregular menstruation, or cramping throughout their periods. They use a type of medication to control the menstruation." This is very alarming to what extent taking this particular medicine is safe. There is space for thinking about this matter. Saline water creates different problems for women. Indigenous woman 6 said, "I get skin diseases from the water, and I have more problems when I get my period. Since sickness can quickly spread to the vaginal areas, women working in the saline water are more exposed to this kind of sickness." The women have issues regarding maintaining personal hygiene. Indigenous women-9 said, "The lack of decent restrooms here is a major issue for us. It gets worse when a calamity strikes, in particular." Also, medicines and sanitary pads are not seen to be used by the women in that area. Indigenous woman 1 said, "We do not have money to buy nutritious food, how is it possible that we will buy sanitary pads? They cost 100 Taka for 8 pads. This is really expensive." Therefore the health condition of the Indigenous women is not so good. Also, the treatment procedure and maintaining personal hygiene is like a luxury there.

Low Education and Early Marriage

In the coastal areas, the Indigenous people do not send their children to schools. Only a small part of them send them to school; rest of the community people send their children to work in the nearby areas. Poverty is the main reason behind this. The Indigenous woman 6 said, "We do not have the ability to send our children outside for further education." If a child passes the primary level education, they are not likely to send their children to secondary schools. Also, the secondary schools are very far from their villages. Indigenous woman 5 told us, "The secondary schools and colleges are very far from our areas. We need transportation costs for that. We cannot pay the tuition, how we will be able to provide them extra money?" This is really challenging to get the children educated about the Munda Indigenous community. For the Munda Indigenous girls, the story is more painful. They get married at a very early age, which is not so healthy. Munda Indigenous woman 8 said, "Parents rush to get their daughter married if she passes the SSC (Secondary School Certificate) exam." Early marriage of girls is an escape route for defeating poverty. Thinking like this in the 21st century is really terrible. They need to change this, and for the change they need education.

Social Stigma

Social barriers exist in the Indigenous communities in the coastal areas. They need to seek permission from the elderly male of the family. And, if they go out for work or do something better, people do not take it positively because of their jealousy and negative attitude. Munda Indigenous woman 1 said, "We don't have any freedom both in the family and in the society. Suppose we go out of our home after the evening people will gossip about that. It's not about my community but everybody does this." They do not have permission to stay out home after the evening. Another common practice there is spreading rumors without knowing the fact. Mostly the women of the community do this. Indigenous woman 1 added, "If I speak with someone for a long time, the other person will say what they are speaking about. They would never ask us what you are discussing. And they will spread rumors." These people never realize the need to go out, educate, and communicate with the outside world. Thus, they still are living in an era where society is going backward.

Inequality in Work Place

The Munda Indigenous women were frustrated when we were listening to their stories about the existing inequality in the workplace. Indigenous woman 4 said, "Two hundred Taka is given to the women and 300 Taka to the men. I also work. However, the work isn't always accessible. Even though there are a lot of people looking for jobs, there isn't much available." Although there is a crisis of jobs, Munda women are not paid as equally as men. They are not chosen for a better job even if some of them are educated. Still, they believe and pray for a better payment. Women are bound to work in exchange for lower wages. Regarding this issue, Indigenous Woman 10 mentioned, "I desire that we will be paid equally. But we do not leave our jobs after this differentiation. Because if we do not go to work someone else will go and grab the work." Another thing they mentioned when we were listening to the story. It is not like they are working any less being a woman. Indigenous woman 9 mentioned, "I feel very bad and sorry for us. While we all labor similarly, our incomes are lower than those of men. They won't listen to us if we apprise them." Thus, the system there is built in such a way that the labor market is cheap and there are so many people to work. Also, work opportunities are less. This trap makes them work for lower wages.

No communications with Policymakers

When we asked the women about their connection with the local political leaders and the policymakers, we did not get anything remarkable. They could not build any significant relationship with them. But in their stories, they are the most vulnerable and from the ignorance of the government, they are not doing well. Indigenous woman 7 said, "We are the smallest indigenous people in Bangladesh. But we are staying so backward, it makes me sad." Regarding the local leaders we have found that they barely know about the community and they go to the area before any election to seek votes for them. Munda Indigenous woman 9 said, "When the member or the chairman comes here, they keep promising us that they will help us, but they forget their promise after the election." They could let their problems know to the leaders if they had connections with them. Indigenous woman 10 said, "We cannot do anything because we don't have any connection to the outside world." Also, the mainstream does not allow them to go to their office. In this way, they are always left behind. But having a significant relationship with them could be the most beneficial way for them.

Community-Led Solutions from Indigenous Women

Women as Change Makers

We listened to the stories of women who came forward from the barriers of the family and society and became successful. Some women have their own earning source, and some have changed people's views toward women. They contribute to their community significantly in many ways. Indigenous woman 1 said, "I am motivated to learn. I believe that if I can accomplish something positive, I can resolve our issues." Mostly these self-motivated women are changing their community. But these activities are not easy to initiate. Indigenous woman 4 mentioned, "For girls, nothing is simple." Still, these poor Indigenous community girls go to school and take training. Regarding the training Indigenous woman 1 said, "I have got training from the animal department. I became successful. If we utilize the training, we will be successful. But people never try. We get training and after going to home we put those things in the corner." Educated Indigenous women have suggested that other women should utilize every opportunity they get. They need to be more active if they want to change their fate. The women, who have become successful, should be role models for the community. The Indigenous women can follow the leaders and get motivated by their deeds.

Education, Indigenous Women Leadership, and Participation

Munda Indigenous women have made great contributions to their families and society. They mentioned that they can make decisions in the family as their male counterparts do. Munda Indigenous woman 1 said, "We make decisions and plan out our tasks together." They are also interested in the local politics. Regarding this Indigenous woman 6 said, "I believe that I will be able to complete my education. My goal would be to become the chairman or member." She added, "I'll make an effort to raise community awareness and offer guidance to the community. And they will decide what is good for them and what is not." They have the realization that if they can give the lead for their community they will come forward. Their story tells a lot about the strength, patience, and empathy for defeating poverty. Munda Indigenous woman 9 said, "I will complete my study and stand with the underprivileged. For us to advance, I shall consider the Munda community's progress. I'll consider myself successful if I can accomplish these." However, other women have different opinions. Indigenous woman 6 said, "Without education, we cannot enter local politics or hold power. Boys as well as girls need compulsory education till the higher secondary level. If I get the opportunity, I will make education compulsory for the young generation." She added, "I would open a school in my community. I would make an effort to manage fresh water for everyday usage, create a pond, and install a deep tube well for drinkable water." It is really beneficial for the Munda community if they get in the leading position. They can bring a change in their life.

Land Rights and Women Rights

Munda Indigenous community suffers most because they do not have any permanent place to live and they do not have land rights. Indigenous woman 7 said, "I do not possess any land on which I will make a house. I am currently residing in my uncle's house. He would seldom ask me to leave the house." The system made them deprived of their basic human rights. Munda Indigenous women suggested solutions for their rights. Indigenous woman 5 said, "The prime issue is that we have nothing. If we had good jobs and help from the government for a piece of land we could make a living. We want land rights." Women also struggle for their freedom and rights. Indigenous woman 6 said, "Women are asked questions whenever they want to do something. This is not freedom. Women are adults and they can make the decisions. Society needs to be aware of this. Change of thought is a need now." It is crucial to raise community awareness and educate people about crisis response and the value of group support, especially among women. It is critical to alter this paradigm in the coastal areas, where women encounter barriers to participating in voluntary and social activities. Their engagement outside the home can be facilitated by providing them with knowledge on disaster response and encouraging a spirit of mutual aid.

Awareness Building within and from the Community

Usually, mainstream people have a negative view of the Indigenous community. The Indigenous people feel low for this kind of differentiation. Indigenous woman 5 said, "Since we are lower caste individuals, mainstream people are not friendly with us. They ask us to keep a safe distance from them when we get to work. They restrict us from touching their staff." The women of the community think that if they could communicate with the outside world, it would be better for them. Because being an outcasts they are excluded from various facilities. The Indigenous woman 9 said, "The adjacent villages, home to the majority of the population, have an excellent road, but we do not. However, the chairman keeps saying they would assist us, but they never keep their promises." Awareness is needed in the community as well as out of the community. People need to respect each other's belief systems. Indigenous woman 7 said, "We require a safe workplace, reduced salinity, a healthy atmosphere, and increased public awareness. So, we can live in peace together." Mainstream communities or the outer world should learn about the situation of the Indigenous communities for a better understanding.

Language and Cultural Rights

Munda Indigenous community has its own language. Lately, very few people are speaking the language. Indigenous woman 8 said, "Not too many people speak our own language. We are seeing a rise in the use of Bangla. Youngsters are unable to speak their own tongue. However, we wish to communicate in our own tongue." The Munda Indigenous people believe that if they can preserve their language, it will help their community. Munda Indigenous woman 9 said, "We are Bangladesh's smallest Indigenous people. But I find it unfortunate that we are continuing to go backward. Preserving our language and culture will be a way to help the community." If they are recognized for their language, the outer world will know about them. Mutual concern from every community and sector would help in making a better policy for their community. The Indigenous women intend to learn technical knowledge so that they can earn and make themselves better. Indigenous woman 1 mentioned her wish saying, "It would be great if we could produce our handicrafts and the government assisted us in getting trained and marketing the products outside. We represented our culture through the crafts and preserved them." Therefore, they need help from the government.

Discussion and Conclusion

Tropical cyclones, storm surges, droughts, saline invasions, riverbank and coastal erosion, flash floods, and storms are among the human-created environmental disasters that might affect Bangladesh's coastal regions. Climate change has a wide range of negative repercussions on people, animals, and the ecosystem in addition to poverty, uncontrollably low food and water supplies, unemployment, saline incursion, and health problems. The researcher has discussed the women's perspective of climate disaster, the impact of climate disaster, and community-led solutions described by the women of the Munda Indigenous group in this chapter. The discussion and conclusion have three main matters to be discussed, including Rethinking Climate Disaster Adaptation Policy, Indigenous Women's Leadership and Land Rights.

Rethinking Climate Disaster Adaptation Policy

According to our research, the Munda Indigenous community members have very little contact with the local political elite and decision-makers. And for this reason, enacting new laws to address climate change is insufficient. These days, it's normal to see people acting inappropriately and ignorantly toward Indigenous populations. We must alter these policies and mentality. Comprehensive disaster management, flood control, enhancing the resilience of vulnerable groups in rural and urban areas, migration, and other significant issues also need to be addressed. Innovative, open, and comprehensible solutions are needed to solve these problems for the poor (Shammin et al., 2022). To address the particular vulnerabilities of Munda Indigenous communities, the government of Bangladesh must reconsider the policies about climate change adaptation, taking into account the interplay between cultural practices and environmental changes (Garai et al., 2022). Policy responses to climate change should better include the land-based perspectives of the Munda Indigenous community to be in compliance with traditional knowledge and adopt community resilience. It is crucial to incorporate inclusivity into adaptation plans in order to protect the rights and welfare of Munda Indigenous groups in Bangladesh that are experiencing increasing effects from climate change. Climate responses, risk reduction, and mitigation require certain enabling conditions, including transparent multilevel and inclusive governance, enhanced institutional capacity, affordable financial accessibility, and behavioral change (Niamir & Pachauri, 2023).

Indigenous Women Leadership

Munda Indigenous women have made great contributions to their families and society. The research findings reveal that the efforts of the women during a disaster have no limit and are unacknowledged. The administration also did not send any immediate help during the extreme conditions. To provide prompt response and assistance during emergencies, effective leadership and efficient organizations are crucial. According to Crosweller and Tschakert (2020), government resilience policy frameworks include tactics to directly address the socioeconomic drivers of injustice, inequality, and vulnerability. It is important to teach women how to prepare for disasters, how to respond to the disasters either alone or with their children, and how their resilience can help to mitigate the effects of future disasters (Karistie et al., 2023), because Indigenous women often lack leadership roles in disaster-resilient development, increasing vulnerability to threats and affecting their livelihoods, social status, and economic status (Seleka, 2017). Our study also found that because Indigenous women have knowledge of resilience from their traditional roles, they have the potential to lead during disasters. Also, forming partnerships with local leaders, residents, and policymakers at the local, state, national, and international levels is necessary to effectively address environmental concerns, as well as the social, political, and economic frameworks and structures that have shaped and created the environmental impacts that communities face (Austin, 2004).

Land Rights

The study demonstrates that the Munda Indigenous community suffers the most as a result of their lack of land rights and permanent residence. Neither their land nor their rights to it belong to the Munda Indigenous community. They are forced to spend a lot of time away from home in order to look for job because their homes are untenable. In addition to having fewer job options, many of them are forced to go without food, drink, and employment. A lot of them claimed the government disregarded their fundamental human rights and they had no guarantee of a future (Datta & Kairy, 2024). Their lives have been left stranded by poverty, unemployment, salinity, water crises, and other human-created environmental disasters. Furthermore, their lives have been made hopeless by inequalities, marginalization, unfair access to various facilities, and social injustice. Their land rights should be ensured so that they can practice their cultural celebrations. Also, they will be able to preserve their language. Being the smallest Indigenous community, they deserve to have land rights along with other basic human rights and preserve their culture. The Munda Indigenous women also will be empowered once they have their land rights. Bangladesh may respond to climate change promptly by leveraging the land-based knowledge and practices of the Munda Indigenous tribes. Effective strategies to address climate concerns can be developed by taking into account the relationship between the people and the land. Bangladesh can also gain from the wisdom of the Munda people and collaborate for a more promising and resilient future.

7 The Role of Indigenous Laitu Khyeng Youth in Building Community Resilience-Perspective from CHT in Bangladesh

This chapter explores the Laitu Khyeng[1] Indigenous communities' youth land-based perspectives in dealing with their community disaster in Chittagong Hill Tracts (CHTs), Bangladesh.[2] Through story-writing initiatives, this chapter takes a youth-led approach to engage them to investigate the probable causes of the flash floods, take adaptive measures to build community resilience, and propose mitigative solutions for the governments. The chapter follows a decolonial approach where the findings suggest that the innovative and the most effective solutions come from the community within. By deconstructing prevailing discourses and advocating Indigenous youth voices, the chapter elucidates how the confluence of historical injustices, land dispossession, and sociopolitical marginalization exacerbates the impact of disasters on the Khyeng Indigenous communities. The findings of this chapter will advocate for the youth voice in sustainable land management, empowering youth in decision-making to preserve community interest. The findings explain how the government should adopt sustainable development approaches through recognizing Indigenous land rights in the Indigenous areas.

The Laitu Khyeng Indigenous community is a small Indigenous community living in the Bandarban District at CHTs, Bangladesh, comprising 2,459 men and 2,367 women with a total of 4,826 people (Population and Housing Census, 2022). The Laitu Khyeng (referred to as Khyeng in the text) community traditionally relied on the natural resources (water, forests, and lands) as a source of food and livelihood (Adnan, 2004). The Laitu Khyeng community has been living in this region for the past 300 years. The rich ecosystem of this CHT region has been the sole source of food, income, and livelihood for the community (Roy, 2000b).

Various outsiders, mostly from the mainstream community, are involved in activities such as land grabbing and deforestation coupled with unplanned land and natural resource management in the name of "so called development," exacerbating the environmental degradation such as landslides, flash floods, droughts, and frequent rainfall (Islam & Hyakumura, 2021). These extreme weather events have intensified the vulnerability of the Laitu Khyeng community people. Years of sociopolitical oppression and ethnic cleansing by the government and local authorities have created a mistrust among the

DOI: 10.4324/9781003536345-7

Khyeng community (Roy, 1998). The socioeconomic needs of the Laitu Khyeng Indigenous community are being ignored and deprioritized by the local government for a long time, which has reduced the representation of the Khyeng community in decision-making, and in accessing and managing their traditional land and natural resources sustainably (Datta, 2015).

Traditionally, the Khyeng have developed and practiced mechanisms of extracting natural resources in a sustainable manner. Their sustainable land and forest management techniques were significant to adapt to climate events and to build community resilience (Datta, 2018a). The Indigenous land-based knowledge is passed down to the next generations through storytelling and evidence-based practicing where the youth have a mediator role in passing down their land-based knowledge from generation after generation (Datta 2015; Adnan, 2004). Not only that, the youth have an active role in regaining the lost land-based knowledge, preserving the existing land-based practice and knowledge, and applying it in building community resilience. The Eurocentric education system contributes to the gradual decline of the Indigenous values and traditional land-based practice and knowledge from the Laitu Khyeng youth (Datta, 2015). Due to forced cultural assimilation by the outsiders, young generations of Khyeng have been separated from their traditional land-based practices.

Indigenous communities in Bangladesh have received limited research attention (Rahman et al., 2023), and they are represented in a way that they are not welcoming, poor, illiterate, cynical about Westernization, and hold a backward mentality. However, a few numbers of research are found on the Laitu Khyeng community. The traditional literature has purposefully ignored the systematic marginalization of the Indigenous people in CHTs (Roy, 2000a). Banerjee (2000) stated that mainstream research to a great extent justified the grabbing of traditional land in the name of development.

The Laitu Khyeng have an age-old land-based practice that is deeply connected with their daily lives and is providing the sources of life and livelihood. For centuries, they have acted as the protector of lands and forests as they have developed sustainable techniques for extracting resources without disturbing the ecological balance (Adnan, 2004; Mohsin, 2002; Roy, 2000a). These sustainable land-based practices have passed down to generations through their spiritual connection with land, their identity, their culture of storytelling, dancing, singing practices where the elderly act as the knowledge keeper and the youth act as the learner (Datta, 2015). These invaluable land-based knowledge, values, and practices are diminishing due to the government's disproportionate emphasis on the colonial education system (Datta, 2018a; Shaffril et al., 2020). It is evident that government's discriminative environmental management policies through declaring reserve forests, empowering settlers to practice illegal land grabbing, brick field, monoculture, tourism construction, etc. has a severe consequence on environmental damage, fostering climate change, and on the life and livelihood of Khyeng community (Datta, 2015). The continuous deforestation in the name of development work has reduced the number of local species, loosened the hilltop soil, silted the riverbed, and contributed to the prolonged flash floods. In August 2023, the Khyeng experienced one such devastating prolonged flash flood. All of these anthropogenic activities are making the Khyeng community more vulnerable to extreme climate events and are reducing their capacity to respond. Therefore, in this chapter our goals are to learn the Laitu Khyeng Indigenous communities' youth land-based perspectives in addressing community disasters, utilizing storywriting initiatives to engage youth in investigating the causes of flash floods, adaptive measures for resilience, and mitigative solutions for governments. Using a decolonial land-based approach and highlighting Indigenous youth voices, the chapter reveals how historical injustices, land dispossession, and sociopolitical marginalization intensify disaster impacts on the Khyeng Indigenous communities and advocates for youth involvement in sustainable land management and decision-making to preserve community interests and promote government adoption of sustainable development through power redistribution and mutual respect.

Theoretical Framework and Methodology

In this research, we used both decolonial and land-based theoretical frameworks to challenge existing paradigms that often marginalize or misrepresent Indigenous perspectives (Datta, 2023, 2018). Current research methods are inadequate to accommodate land-based knowledge (Wheeler, 2001). Using the land-based framework, we seek to decolonize both we as researchers and our research according to communities' needs, challenge dominant narratives, and explore ways to reclaim the meanings of research from and within Indigenous community perspectives. This approach recognizes the inherent bias within traditional research methodologies and strives to break free from Eurocentric perspectives, fostering a more equitable and respectful engagement with Indigenous knowledge systems (Smith, 2018). The land-based relational theoretical framework aligns with the principles of Indigenous self-determination and resurgence (Kovach, 2016). It emphasizes the importance of centering land-based values, knowledge, and experiences within the research process. This framework also believes that the solutions should come within the community, and it should respect the value of the land-based culture and promote and support land-based knowledge (Wheeler, 2001). By grounding the study in youth-led principles, we aim to be "the voice" of the community without forsaking the ancestor's knowledge. This framework enables a more holistic understanding of land-based issues, acknowledging the interplay between cultural, social, and environmental factors. Ultimately, the land-based theoretical frameworks serve as a commitment to respect, collaboration, and obligation with land (Datta, 2018).

This field research was conducted in August 2023; we used story writing and deep listening methods to facilitate knowledge exchange and learning. In addition to the 10 Khyeng youth, 10 Khyeng Indigenous elders guided in deep conversations, sharing traditional knowledge that has guided their survival during flash floods. Through these collective conversations, the Khyeng Indigenous youth created reflective writings on the causes of climate change and disasters, primarily attributing them to various undesirable projects in their community and explored ways to challenge and rebuild environmental sustainability. These youth reflections are in the process of being published in a youth-led journal at the UNDP. The elders recommended that researchers highlight youth perspectives in academic publications to inform international academia and policymakers. Thus, this chapter serves not only as our research but also as our responsibility to highlight Indigenous community needs, ensuring we remain respectful of the Khyeng community's culture and practices. In addition to following institutional research ethics, we strongly followed Khyeng communities' traditional protocols and continuous forms of consent. We tried to center their voice and needs in our findings.

In analyzing our research findings, we followed community elders' suggestion on focusing main themes, particularly based on community needs. We learned three main themes (Table 7.1): climate change and its impact on the Khyeng community, the Khyeng community's land-based practice to adapt to flash floods and how the settler's activity is endangering the Khyeng

Table 7.1 Themes and subthemes of this chapter

Theme	*Subtheme*	*Examples*
Climate change and its impact on the Laitu Khyeng community	– Flash flood – Impacts of flash flood in life and livelihoods of the Laitu Khyeng	Loss of lives, destruction of houses, cattle etc., cut off from the livelihood facilities, disruption of transportation, shut down of schools, landslides, loss of crops.
	– Settlers' invasion is prompting extreme weather events such as flash floods	Brick field, monoculture, declaring reserve forests, unplanned development structure, illegal land and river grabbing
	– Youth perception of flash flood	… "when the river siltation happens due to environmental damage, the rivers overflow water and causes prolonged floods and heavy loss of lives and livelihoods."
Land-based practice in building community resilience	– Youth led proposed measures to build community resilience to adapt with flash floods	Protecting and promoting Khyeng language, incorporating Khyeng land-based practice in the mainstream curriculum
Role of the government to include the Khyeng youth in the disaster management	– Youth-led approach in restoring and protecting Khyeng land-based knowledge	Recognize the Khyeng land-based practice and its implication in integrated flood management, ensure dignity and equal participation of Khyeng people in designing curriculum, redistribution of land, allocate research budget

community, and the role of government in including land-based knowledge in mainstream education. Table 7.1 illustrates the themes and subthemes of this research:

Findings: Laitu Khyeng Land-Based Practice in Building Community-Level Climate Resilience

Khyeng community's life and livelihoods are deeply affected by climate hazards such as flash floods. It is evident that settlers' activities such as brick fields, monoculture, reserve forests, land and river grabbing are fostering these climate hazards. Under this context, the revival of Khyeng land-based knowledge connecting the youth can be a sustainable way to combat climate change. However, to achieve that recognizing the Khyeng land-based knowledge, incorporating their land-based practice in the mainstream curriculum and ensuring equal participation of the Khyeng in policy formulation are the need of the hour.

Institutionalization of Khyeng land-based practice is a must to revive their traditional knowledge and to implement this knowledge in building community resilience. However, the Khyeng youth are far away from their traditional land-based practice as the mainstream educational curriculum is diluting their land-based practices and disconnecting them from their roots.

We took a youth-led approach, where the Khyeng youth engaged in story writing on identifying the cases of flash floods, and their impact on the community and proposed solutions from the perspective of Khyeng land-based practice. They have identified deforestation, monoculture, and brick industry as the biggest contributors to environmental degradation that are causing prolonged flash floods. They have also proposed short-term and long-term suggestions in integrated flood management considering protection, preparedness, and recovery perspectives. Some of the long-term measures include cultivating traditional plants, promoting *jhum* cultivation, and integrating land-based knowledge into the education curriculum as long-term measures.

The Khyeng youth suggested that the existing educational curriculum should recognize and include land-based traditional knowledge and its implication in sustainable resource management and climate resilience. They have recommended that the land rights should be redistributed to the Khyeng community, the development projects should ensure equal participation, and research funds should be allocated for creating a national database to preserve Khyeng land-based knowledge.

Meaning of Climate Change to the Khyeng Youth

In August 2023, the maximum precipitation was recorded at 165 mm in Bandarban District (*The Daily Star*, 2023). The heavy rainfall caused flash floods and landslides. While we were doing the participatory discussion and deep listening, the Khyeng people said that during August 5 to 10, 2023, they experienced the devastating flash flood in the history ever. Although the Khyeng community was affected severely due to the flood, there was no initiative from the government to estimate the loss and damage. Some of the nongovernmental organizations did the assessment for their internal use, which excludes the Indigenous community. Need Assessment Working Group hosted by CARE Bangladesh in its assessment report estimated that 1.2 million people and 34,973.5 hectares of cropland in the CHTs were affected due to the flash flood.

The recent flood affected the Khyeng youth physically and mentally. Most of the Khyeng students suffered from fever and stomach diseases due to polluted drinking water as fresh water was scarce during floods. Schools were shut down for weeks, which affected their overall education. After the opening of schools, they will have to cover the syllabus within a very limited time that will create huge physical and mental stress. Em Prue Khyeng, a student of ninth grade, said, "All of our books got wet due to the flood, we have to buy new books now and our parents have to bear extra cost for that." During the story-writing process with the Khyeng youth, they shared their perspective on climate change and its impact on their community, how human-induced activities are inducing vulnerability for them, and their role in mitigating the disasters. Chinoprue Khyeng, a student of 10th grade, said, "Climate change means snow melt in the higher reaches, heavy rainfall in the lower course, frequent tropical cyclones and floods in the hilly areas of Bangladesh." Uashaiprue Khyeng, a student in 9th grade, said, "Bangladesh is abundant with river flows, and numerous people's lives and livelihoods depend on the river. But when the river situation happens due to environmental damage, the rivers overflow water and cause floods and heavy loss of lives and livelihoods." Searching Khyeng, another student, expressed, "This is the most prolonged flood I have ever seen, the water level went as high as our rooftop, our house washed away, and we had to stay at the roads under the sky."

The Khyeng community's vulnerability to climate change is increasing as a result of the government's invasion of their traditional lands and declaration as reserve forests, which has empowered the settlers to exercise unlawful control over them. Unplanned development projects, rubber plantations, and other such activities have further increased this vulnerability. During the story sharing an Khyeng elder said, "Our forefathers are the owner of this land, no one knows the land better than us. We have learned to use this land without destroying the ecological balance. Our land management practice is destroyed by the settlers activities."

Flash Flood in 2023 and Its Impact on the Khyeng Community

The evidence suggests that the long-time ignorance from the state directly or indirectly prompts the environmental degradation in the form of unplanned construction and land grabbing, causing river siltation and making the Khyeng community more vulnerable toward climate hazards such as flash floods. Ngo Jai U Khyeng said:

> No organizations came to help/support us the post-flood loss and damage. The settlers from the nearby community have received a sufficient amount of relief, and financial support to rebuild their home, but we did not received anything. We have received nothing from the government although our houses are damaged fully.

The flash flood has a severe long-term impact on the economy and livelihoods of the Khyeng community as it causes scarcity of food, and damaged crop cultivation with heavy alteration in basic health, hygiene, and sanitation activities.

Food Scarcity

During the deep listening we found that some of the Khyeng community people had to starve for a week, and some of them lived on steamed bananas as their only meal. The roads and networks were disrupted for a long time. The untimely flood damaged all of their crops and saplings and doubled their cost of crop production. The transport and communication were shut down completely. The Khyeng community was disconnected from the rest of the world for more than a week. Since it was raining heavily with the water level rising above the waist, it was even difficult to communicate within the community. "The flash flood was so sudden that we had no preparation. We had to starve for days. Since the communication was cut off, we could not even go out to buy food," said Ngo Jai U Khyeng.

Housing and Shelter Crisis

Most of the Khyeng participants' houses and cattle were washed away during the flood. The Khyeng had no place to take shelter as they had no access to the cyclone shelter. Most of the Khyeng families had to live under the open sky. The timing of the rain was very unusual, and that's why none of them were prepared for the sudden flood. An elder Khyeng said, "The duration and the precipitation rate has changed drastically over the time that has worsened our condition." From the discussion with the Khyeng participants, it was evident that having a place in the cyclone shelter and receiving relief is highly subject to political connection. Due to not having political connection, the Khyeng community does not have the access to cyclone shelter. Settlers are given priority in accessing shelter and receiving relief. "It feels like we are the second or even third category citizen in Bangladesh," said the Khyeng elder during the learning exchange discussion. The water level was so high as ever that most of the Khyeng people had to live on the dam under the open sky, and only some of them were lucky enough to take a temporary shelter in their relatives' houses.

They received 5 kg rice, 250 ml oil, 1 kg of lentils as relief, which was insufficient to survive the flood, whereas the settlers received a good amount of relief to survive. The words from Ngo Jai U Khyeng said it all, "It is a mockery in the name of relief."

Scarcity of Drinking Water

The Khyeng community is experiencing water scarcity due to ongoing cash crop production such as rubber, teak tree, tobacco cultivation, and brick field has contributed to severe deforestation in the CHT area. The Khyeng community stated that these cash crops tend to consume more water than the native crops, which contributes to lowering groundwater levels. As a result, the streams (*jhiri*), the only source of fresh water for the Khyeng, are drying gradually, and the community is facing drinking water scarcity throughout the year. The situation worsened during the flash flood. They had to drink rainwater during the flash floods that caused waterborne diseases. "We drink rainwater as only rainwater is available during the flood, as a result, children and elders suffer stomach diseases" (Julia Khyeng, social worker).

Health and Sanitation Crisis

The flood in 2023 caused severe health crisis during which people were struggling for the basic facilities such as food and drinking water, and experienced sanitation and hygiene crisis. The sanitation facilities were fully damaged, water sources were contaminated, and tube wells were fully non-functional during that flood. Khyeng women and girls faced severe sanitation and hygiene crises since women and girls have special needs for sanitation and hygiene. Stomach diseases and viral fever are very common in the Khyeng community after any flash flood especially among children. Julia Khyeng said that women and children did not receive any medical support from hospitals or clinics since the roads and transportation were shut down completely due to the flood. They had to buy medicine with double or even triple prices from the local market.

Loss of Crops, Loss of Sources of Income

The Khyeng community heavily rely on agriculture for their livelihoods. Crops, seeds, and saplings are damaged heavily due to the sudden floods. The flash flood has a long impact on the cultivation process. The sudden flood in 2023 washed away all the saplings of the farmers, and they had to buy saplings at a higher price from the market. As most of the Khyeng are poor farmers and day labors, they are heavily dependent on the local money lenders (*mohajon*) to buy seeds, fertilizers, saplings, etc. Oftentimes the local money lenders took advantage of the situation by claiming higher interest rate and the Khyeng had to accept that as there were no other sources of formal or informal finance. The

Khyeng farmers are sad that the price hike of the agri-inputs and the higher interest rate will increase the cost of agricultural products and thus will reduce their income.

Due to unusually prolonged floods in August 2023, the Khyeng farmers had missed the sapling plantation season, which poses a risk of lesser crop production. The Khyeng farmers anticipate that the crop loss will increase the credit burden on them.

An elder Khyeng said that:

> There is no organization to give us loans, we have to take loans from local lenders (*Mohajon*) with high interest. Although the interest rate is high but we have no choice because we need money for crop production and rebuild our damaged houses after the flood. We have to work even longer time to repay the stressful loan.

Land Slides

Frequent landslides at the time of heavy rain and flood are a death trap for the Khyeng community. All members of three families died due to landslides in the recent flood. Rubber and teak tree plantations, unplanned tourism infrastructure, tobacco cultivation, and brick fields are the main causes of landslides. Most of these settlers' activities are controlled either by the government or by the local political people who hold the power. Julia Khyeng, the Khyeng community social worker, said, "We are very afraid of landslides. Landslides is becoming more frequent and dangerous for us as we live on the foot of the hills. The soil of the upper hill area is used for brick fields and is the biggest cause of landslides." The flash flood has a long-lasting impact on the lives and livelihoods of the Khyeng community, including food and shelter crises, health and sanitation crises, losses of crops, and slides. However, anthropogenic activity-induced environmental degradation has worsened the situation by increasing the severity and frequency of the flash flood.

Causes of Frequent Flash Floods

The Khyeng youth have stated that their ancestors' lands are being grabbed in the name of development of the community, but in the end, the community were deprived of from the control and access over land and water. This continuous suppression has made them vulnerable to frequent disasters and diminished their ability to fight against climate change. With an increasing number of brick bricks, tobacco, *gamari* and teak tree plantations, and tourism construction, the Khyeng are losing their lands. The rich biodiversity of Bandarban District, which used to be sustainably managed by the Khyeng community, is now under threat due to declaration of reserve forests, systematic land grabbing, and unplanned development works. The Khyeng's age-old land-based practice deeply connected to land and water is now under extinction, which endangers the Khyeng community. The government should respect the traditional land-based practice and provide decision-making power to the Khyeng community while designing land management activities.

During the youth-led story-writing session, the Khyeng youth identified some of the prominent reasons for the flash flood in CHTs and proposed some innovative solutions using land-based practice to promote community-level disaster sustainability. The major causes of flash floods identified by the Khyeng youth are:

Massive Deforestation and Unplanned Development Work

The Khyeng youth identified frequent deforestation as one of the major reasons for the prolonged flash flood. The soil of the deforested area is becoming loose and is being easily slided with heavy rainfall, which is causing river siltation. On top of that, river grabbing by the settlers and construction development works on the bank of the river are creating river siltation and impeding the natural flow of river. All these anthropogenic activities are reducing the capacity of the river to absorb rain water, and thus, the rivers are overflowing frequently and causing flash flood. The duration of floods is also increasing since the capacity of the rivers to absorb rainwater has decreased drastically due to unplanned development activities.

Popularization of Cash Crops

The government-backed cash crop cultivation (*rubber, tobacco, teak tree, gamari*) replacing the natural forests is a major concern for the Khyeng community. The first teak plantation started in Bangladesh in the CHT area in 1871 (Hossain et al., 2008). Now, almost 70% of the cultivable area in CHT is occupied by teak plantation, which is a greater threat to the ecology (Feroz et al., 2014). This monoculture of cash crops is destroying the traditional forests and their related spirituality. With this monoculture, the soil fertility and moisture content are decreasing due to the continuous cutting down of local trees and replacing them with teak, tobacco, *gamari* trees (Mahmud et al., 2017). The Khyeng youth emphasized that the hilltop soil containing less mineral and less moisture is easy to deplete in the rainy season causing river siltation and contributing to river overflow.

Brick Field

The Khyeng youth believe that the brick fields are the silent killer of the biodiversity in the CHT, destroying the ecological balance and is contributing to flash flood. The youth also believe that brick fields are evidence of power exercise and oppression by the outside Bengali settlers toward the Indigenous people. Many times the brick field owner forcefully throws out the Khyeng from their lands, and the government is reluctant to take any actions on that. Every year more than hundreds of brick fields are built in the CHT. Air pollution and carbon emissions from the brick field are one of the major causes of environmental damage in the CHT (Datta, 2019). The brick field requires a huge place to operate and timbers as fuel; massive deforestation in the hill area is occurring to run these brick fields and contributing to landslides and flash floods. Government invasion, illegal power exercise on the Khyeng community, land grabbing, and unsustainable development projects are the biggest reasons for biodiversity loss in the CHT region that contribute to climate events such as flash floods and heavy rain. Massive deforestation, cash crop cultivation, and brick fields are some of the detrimental projects taken by the settlers that have destroyed the traditional forests of the Khyeng, ruined the sources of their livelihoods, and marginalized them to extinction.

Adaptive Measures to Build Community Resilience

The Khyeng community's land-based techniques and community bonding are proven to adapt to climate change. For example, an elder Khyeng said that boats made using their own traditional techniques are very helpful at the time of communication and transportation when roads are shut down due to flood. The food and seed storage techniques of the Khyeng are also a critical aspect of disaster management. However, the formal education system is excluding land-based knowledge from their curriculum and hence is failing to design techniques sustainable for community disaster management. With the expansion of modern education, the Khyeng youth are losing ties with nature and with their older generations. The silent but gradual disconnection from land-based knowledge is making the Khyeng community even more exposed to climatic change and extreme events such as flash floods (Datta & Kairy, 2024). However, the youth have a great role to play in preserving the land-based knowledge of the Khyeng in building climate resilience.

Short-term measures: While proposing the solutions, the Khyeng youth emphasized protection, preparedness, and recovery. Protection defines flood protection facilities and infrastructure. Preparedness defines engaging the community and youth to build climate resilience. Recovery defines environmental and socioeconomic restoration. Their proposed short-term adaptation measures include a local language-based early warning system, building elevated houses, and building emergency shelters. Uashaiprue Khyang's perspective is, "Most of the Khyeng people speak their own language. The early warning system will only be effective when it is disseminated in the Khyeng Language. The youth involvement in disaster messaging is also important."

Long-term measures: The Khyeng youth proposed some long-term adaptive measures on community-level disaster preparedness and recovery in light of their age-old land-based practices. Although it is proven that the traditional land-based Khyeng knowledge is critical for sustainable management of natural resources, and disaster preparedness, the mainstream policymakers are yet to explore the full potential of the land-based practice. The Khyeng youth asserted that the government and policymakers should document their land-based knowledge and engage Khyeng youth in adapting community-level climate resilience. The long-term proposed measures include the following.

Promote Indigenous Plants

The Indigenous plants provide benefits such as producing oxygen, storing carbon, regulating air temperature, holding soil from being depleted, and filtering runoff. Many countries have adopted Indigenous plantations on the slope of the hills and riverbanks as a technique for integrating flood management. The Khyeng youth asserted:

> The local plants are the sources of food and livelihoods at time of disasters such as floods. Local plants are good for making boats which is the only means of communication at times of flood. Fruits the local plants were the main sources of food at time of recent flood.

That is why it is the need of the hour to plant Indigenous and local trees instead of short-lived cash crops.

*Promote Traditional (***jhum***) Cultivation*

Small-scale *jhum* cultivation or shifting cultivation practiced by the Khyeng Indigenous is considered beneficial for biodiversity. The seasonal and shifting cultivation practice used to be the main source of livelihood for the Khyeng community people. However, 70% to 80% cultivable lands are now being replaced by various types of profit-making crops. The irony of fate is that the marginal Indigenous communities are being forced and misled by the outsiders to practice monoculture in the name of profit making. However, the major share of the profit is taken by the outsiders, and the community people are left with even more poverty. The Khyeng youth stated that traditional *jhum* cultivation is critical to preventing hilltop soil from being eroded.

Integrating Khyeng Land-Based Education in the Existing Curriculum

The Laitu Khyeng have a rich tradition of respecting nature. They have animal gods, crop gods, land gods, water gods, insect gods that are reflecting on their day-to-day life. They believe that land and water are their teachers. For example, the *jhum* crop symbolizes sustainable management of land, water, and forest (Datta, 2015). Figure 7.1 describes various stages of *jhum* cultivation and its implication in sustainable land management rooted in Khyeng traditional beliefs. Here the *jhum* crop is working as a mediator to transfer the land-based knowledge from generation to generation.

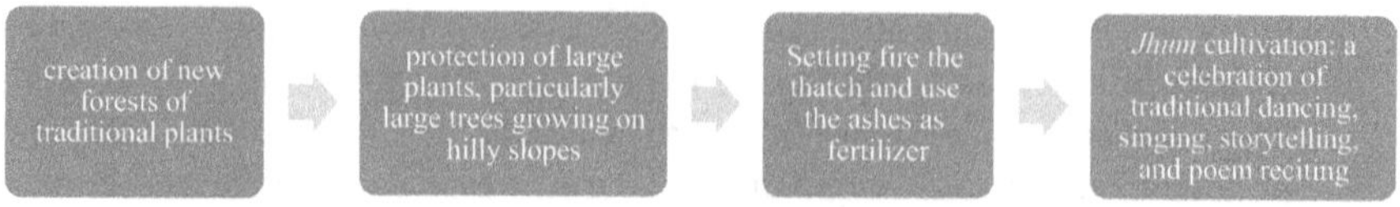

Figure 7.1 Land-based knowledge transformation to next generations via *jhum* cultivation.

With the expansion of modern education, the Khyeng youth are losing ties with nature and with their older generations; thus, they are becoming ignorant about the significance of conservation, and the outcome of environmental degradation, ecosystem loss, etc. (Barua & Ando, 2002). The Laitu Khyeng youth also asserted that they need to be equipped with traditional cultivation knowledge, Indigenous plant identification, and land and water-related spiritual practices. Youth education should reflect a spiritual connection with land, water, and traditional sustainability. That is why it is the need of the time that the education curriculum should reflect traditional land-based practices. The Khyeng youth urged that their land-based practice should be at the center while designing integrated flood management system for the Khyeng community.

Community Recommendations

The forests, lands, and water are interlinked to the life and livelihoods of the Laitu Khyeng community to the extent that it is almost impossible to separate both. Their day-to-day land-based practice and knowledge have traveled through generations in the form of storytelling, dancing, and singing. This age-old land-based practice has inspired the Khyeng community to develop sustainable consumption, protection of biodiversity, and climate resilience. Since the connection between the Laitu Khyeng community and their land water management is so deep, it is not possible to develop an integrated flood management plan without recognizing and integrating their land-based practice. In the course of designing an integrated flood management plan for the Laitu Khyeng community, some of the proposed recommendations for the government and policymakers are given here.

Recognition of the Laitu Khyeng Land-Based Practice

Successful integration of land-based knowledge is largely dependent on the willingness of the government, political leaders, civil society, etc. More research, dialogue, and talks should be organized to acknowledge the importance of the Laitu Khyeng land-based practice. The government should plan for the schools to organize activities and day celebrations where the Khyeng students should be encouraged to talk about their land-based practices, values, and ethics.

Land Rights and Empowerment

Initiatives should be undertaken to mainstream the Khyeng communities by providing them with land rights, decision-making authority, and empowerment opportunities. Ensuring equal rights and opportunities is essential. During youth-led interaction, the Khyeng youth emphasized land rights over relief. Illegal land grabbing, encroachment, brick fields, and monoculture in the Khyeng land should be stopped by the government. Community development projects should prioritize the interests of the Khyeng community, respect their land-based practice and spirituality, and ensure equal power distribution. It is evident that Khyeng community disaster management plan should come from them, because they have a better understanding of the lands and forests.

Incorporating Khyeng Land-Based Knowledge in the Educational Curriculum

The existing homogenous education system is promoting Western ideologies by ignoring the contribution of land-based practice in preserving lands, waterbodies, and forests (Barua, 2004). The imposition of a homogenous education system is alienating the Khyeng youth against their own language. Because language is the vehicle for transferring knowledge, the Khyeng land-based practice is on the verge of extinction since their language is dying. At present, the Khyeng language is one of endangered languages in Bangladesh (*Dhaka Tribune*, 2023). Thus, popularizing the Khyeng language among the Khyeng youth is urgent to keep their land-based knowledge alive. The Khyeng alphabet was created in 2017, and the Khyeng language keyboard on "Hyow" script was launched in 2023 (*Dhaka Tribune*, 2023). It is clearly understood that an increasing representation of Khyeng folks as teachers, advisors, researchers, and community engagement leaders is required to incorporate the Khyeng land-based knowledge in the education curriculum. As of now, the fact that neither the government nor private organizations have any policy to include Khyeng knowledge in the curriculum is a matter of regret.

National Database on Khyeng Land-Based Practice

With deep regret, the Khyeng youth said that "we do not have any national or local level database to preserve our land-based practice and knowledge. Khyeng elders are the last generations of the land-based knowledge keeper. When the elderly people die, the Khyeng knowledge, practice, and culture will also die." The government should prioritize Khyeng knowledge database development on an urgent basis. This will help the policymakers design and implement the right policies to integrate and preserve their knowledge.

Fund Allocation for the Community

The Khyeng community is one of the less funding-allocated communities in Bangladesh. The funding is very limited for land-based research. The government and nongovernment research organizations should allocate funds to promote education and awareness of land-based education, culture, values, practice of land-based conservation and community resilience.

Concluding Remarks

For centuries, the Khyeng land-based practice is working as a tool to manage land, water, and forest in a sustainable way. For centuries, they have acted as the protector of lands and forests as they have developed sustainable techniques of extracting resources without destroying the ecological balance. Their sustainable land-based practice has passed down to generations through their spiritual connection with the land, their identity, and their culture of storytelling, dancing, and singing practices where the elders act as the knowledge keeper and the youth act as the learner.

Ongoing colonial impacts from both governments and development industries have had detrimental effects on the environment that has reduced the number of local species, silted riverbeds, and fostered severe climatic events such as flash floods. Furthermore, the settler's invasion has endangered the life and livelihoods of the Khyeng community people and made them vulnerable toward disasters. For example, in August 2023 the Khyeng community experienced a devastating flash flood ever in the history since 1975 where the human-induced activities contributed to severe losses of lives, loss of crops with an increasing crisis of drinking water, and health, sanitation, and hygiene crises.

In their story writing the Khyeng youth have identified anthropogenic activities such as rapid deforestation, brick fields, and monoculture practices in their traditional forest area as the major contributors of flash floods. Furthermore, they have suggested some measures highlighting their land-based practices that include the planting of traditional plants and *jhum* cultivation that are effective to holding the hilltop soil from being depleted and to providing aids at the time of floods. In the Khyeng community the youth have a great role to play as a knowledge keeper and as a mediator to pass down their land-based knowledge to next generations in building community-level climate resilience. That is why the Khyeng youth stressed that their land-based practice should be incorporated into the formal education curriculum as a tool for sustainable land management and building climate resilience.

Notes

1 There are two clans existing in the Kheyang community: Laitu and Kantu. The Kantu Kheyang live in the hills, whereas the Laitu Kheyang live in the plain land majorly in three villages: Gungru Para, Gungru Mukh Para, and Gunru Modrom Para in the Bandarban District in CHT (Adnan, 2004; Chapola, 2022). It is assumed

that the Kheyang people migrated to the CHT in the 17th century from Myanmar. The Kheyang received 3,000 acres of land during the British colonial period for *jhum* cultivation. Since then, their lives and livelihood are heavily dependent on forests and lands (Datta, 2015). Historically, the Kheyang community has practiced sustainable land management, collective decision-making, and prioritizing community interests as they have a deep relationship with lands that pass through generations. They have developed a deep spiritual connection with nature, and their land-based knowledge management learned from the older generation through informal practices such as storytelling, and sharing, community bonding played a crucial role in land management (Datta, 2015).

2 Historically, the rich, unique and diverse ecosystem of CHT in Bangladesh is the nest of 12 Indigenous communities; the Laitu Kheyang community is one of them (Chittagong Hill Tribes of Bangladesh, 2010). With continuous land grabbing by the government/settlers in the name of administrative and tourism structures, brickfields, tobacco plantations, lumber plantations, and reserve forest projects have diminished their sources of livelihoods (Islam & Hyakumura, 2021). On top of that, the community is being marginalized by years of oppression, increasing levels of power exercised by the settlers and controlling over resources, unplanned development work, sociopolitical subjugation, etc. That limits their capacity to respond to climate change (Datta, 2015).

8 Advancing Indigenous Community-Led Disaster Adaptation

Indigenous community-led disaster adaptation in Bangladesh exists in conjunction with their traditional land-based cultural practices and knowledge. Historically Indigenous people are deeply rooted in their land-based practice and spirituality. These practices have grounded and linked Indigenous communities together and with one another through time. However, today, the space does not exist in isolation, but within the larger context of the regional, state, and colonial structures, Bangladesh's Indigenous communities are situated. Just as the weather and climate within which Bangladesh Indigenous communities exist and are informed by climate change, their adaptation and resilience are also interconnected with these external social, economic, and political structures.

Decolonizing disasters and relearning adaptations from Indigenous land-based perspectives in Bangladesh, the subject matter of this book, is important for addressing climate change. However, to grasp its importance for one such as myself, as a white Canadian settler woman, requires an open mind, and a curiosity for learning the meanings of Indigenous land-based climate adaptions. Also, learn and respect from the Knowledge keepers and Elders within Indigenous land-based communities. Without an open mind, and an open heart, combined with a keen desire to learn, operationalized by deep reflection and consideration and a commitment to "relearn" through a continual process of considering my own biases and prejudices I risk not truly hearing the wisdom within each of the proceeding chapters. I may conflate and diminish the chapters' knowledge perhaps by compartmentalizing what I hear into my pre-existing and established knowledge and understanding.

The Changing Climate in Bangladesh

The significant warming observed all over Asia has "intensified the threat to social and economic sustainability" (Shaw et al., 2022: 1). The recent unrest in Bangladesh can't be directly linked to warming temperatures, recent heatwaves, and climate impacts affecting water, food, and energy supply, exacerbating risks to livelihoods and economy. Water stress has increased due to both climatic and non-climatic drivers, and the future climate risk is

DOI: 10.4324/9781003536345-8

projected to result in a decline in fisheries, aquaculture, and rice production (partly due to increased pests) (Shaw et al., 2022).

People living in informal settlements (already at a high level) are projected to increase with an increase in migration and internal displacement. "In 2019, Bangladesh, China, India and the Philippines each recorded more than 4 million disaster displacements" (Shaw et al., 2022: 1). More exposure to disasters and vulnerability of populations to increasing heatwaves, floods, droughts, and air pollutants result in vector-borne and water-borne diseases, malnutrition, mental disorders, and allergy-related illnesses (Shaw et al., 2022).

In the future, Bangladesh's extreme heat risk measure by days surpassing 35°C is set to increase from roughly 60 days annually to 133 days by midcentury; the number of tropical nights with a minimum temperature greater than 26 is projected to increase from 56 to 133 days by midcentury (World Bank Group, 2024). While in the past 50-year period eastern divisions have experienced less precipitation annually, especially in spring pre-monsoon months, Western divisions have experienced significantly wetter times, especially in post-monsoon fall months. In the future the south is expected to experience a decrease in precipitation (2020–2039) and the north the greatest increase in annual precipitation (2040–2059), especially during the summer monsoon months (World Bank Group, 2024). Although the timing and severity of extreme precipitation anomalies vary by region, Bangladesh is likely to experience more intense precipitation in the future (World Bank Group, 2024). While Bangladesh has already experienced loss, damage, and economic costs as a result of climate change, the anticipated future changes in climate and extreme weather necessitate addressing the gap in current adaptation efforts and needed adaptations to protect the people of Bangladesh and their livelihoods, especially those of Indigenous people.

Methodological Reflections

As interdisciplinary social scientists studying climate change, adaptation, and response to extreme events, we tackle a daunting problem, often with little or no acknowledgment. First, climate change is real, it is happening, and we, the human race, are the cause of the problem (IPCC, 2007). While it is clear that we have emitted greenhouse gases (GHGs) in our industrial and livelihood practices that have caused this problem, how to address the problem to reduce GHGs and adapt to the changing climate is not clear. Many argue that overarching, enduring power structures of capitalism and colonialism are culpable and need to be addressed (van Ryneveld & Islar, 2022; Gonzalez, 2024). Others focus on technologies that can address climate change issues, with little consideration of social change (Alestra et al., 2023). What is agreed is that the individual, social, organizational, business, governmental, and international change required is gargantuan (Tosun, 2022), and even after decades of climate agreements, tangible change remains elusive.

While lists of climate change actions, potential policies, community-based approaches, early warning systems, and structural adaptations are abundant in the mainstream literature, as the dire consequences of climate change are becoming ever more omnipresent and dire, consideration of what has been overlooked deserves reflection. In this author's opinion (Tosun, 2022), considerations of ontology, epistemology, and power relations are often missing. This is partly a result of perhaps a lack of interest (and therefore understanding or caring about these terms) and the conflict that such considerations can give rise to.

In this book, all authors have spent considerable time and dedicated significant space within their chapters to consider their ontology, epistemology, and resulting methodology. Relational land-based research methodology requires such dedication. And although there is no one definition of what an Indigenous research methodology, ontology, or epistemology is, it is very different from a traditional Western lab-based science method of hypothesis testing. Indigenous knowledge is built on the principle that knowledge is relational and shared with all creation, not just interpersonal relationships or research subjects, but the cosmos, animals, plants, and earth (Steinhauer, 2002). This contrasts with Western research wherein an individual searches for knowledge, thereafter gains, and owns (through authorship, etc.) that knowledge.

Western knowledge and science are separated from art and religion, but Indigenous knowledge is often both religious and aesthetic driven by traditional teachings, empirical observations, and revelations where elders play a major central role in having the longest experience and knowledge. Rituals and ceremonial observance are also important for knowledge acquisition and advancement as they enhance knowledge exchange, but also deep reflection and exploration (Ermine, 1995). Land-based learning is a perspective that regards land as the "first teacher, centering relationships" between all humans, nonhumans, and living things; it involves holistic perspectives that include physical, mental, emotional, and spiritual aspects of interactions (Bowra et al., 2021: 135).

Relational land-based research is not an Indigenous "perspective" but a paradigm or fundamental belief that knowledge is relational, can't be owned, and contains obligations. Relations are not just interpersonal, but relationships are with all of creation—the cosmos, animals, plants, and earth and the researcher are answerable to all their relations when doing research (Wilson, 2001). These obligations are that of sharing and building knowledge together with a requisite axiology or ethics (Wilson, 2001, 2003).

The ethics of relational land-based research include Indigenous people controlling, doing, approving Indigenous research; deep respect for individuals and community; recognition of reciprocity and responsibility, making sure research participants feel safe, deep listening, and reflective non-judgment (Wilson, 2003). Further, relational land-based ethics requires a constant awareness and connection between the logic of the mind and feelings of the heart such that both emotional and cognitive experiences are incorporated into all actions.

Acknowledging one's subjectivity is also a constant process, and chapter authors have expressed this in their statements of positionality (Wilson, 2003).

The ontology, epistemology, and axiology therefore inform relational land-based methodology, including those employed by the chapter authors. The land-based nonhuman-centric relational philosophy engenders a relational method. Conversations, storytelling, re-storing, remembering, and sharing tea all become appropriate methods (Kovach, 2009) and those employed by chapter authors. Fundamental to most is the connection to the land, including land-based practices and relations.

Risk, Vulnerability, and Power Relations

Engaging with Indigenous knowledge and land-based practices to address climate change requires nimbleness. Adopting theoretical frameworks, including adaptation, vulnerability, and exposure, is not without issue. Goldman et al. (2018) conclude that adaptation frameworks, tools, and metrics are not relevant or applicable to Indigenous knowledge, nor fit for purpose given the unique and context-based nature of the solution space in which Indigenous peoples exist (Goldman et al., 2018). Leonard (2021) uses the descriptor adaptation "oppression" to describe the engagement of Indigenous peoples in addressing issues of sea level rise using such frameworks; the frameworks were neither created by Indigenous peoples nor created for Indigenous peoples. The Indigenous axiology, or "responsibility-based thinking," describes a duty to other-than-human and human relations not reflected in these frameworks (Sioui & McLeman, 2014). In the event of catastrophic climate change impacts, McGregor (2014) notes that the other-than-humans are capable of surviving without humans and it is the humans who need to learn to live with the land and shifting parameters of climate change (McGregor, 2014).

While the Intergovernmental Panel on Climate Change (IPCC) has recognized that Indigenous knowledge is an important resource for adapting to climate change (IPCC, 2014, 19) and also important in enhancing decision-making and governance (IPCC, 2019), it has only given it the most "thin" recognition, incorporating it into its colonized science institutions including "adaptation." Colonized science, including the inherent "victimhood" of vulnerability (and adaptation) (Nursey-Bray, 2023), is insulting to the depth and breadth of Indigenous knowledge, thus perpetuating a Western view of "science," that is dismissive of Indigenous ways of knowing. Part of the inherent contradiction is that Western ways of developing strategy and facilitating adaptation tend to have a linear logic; by contrast Indigenous ways of doing and being are circular with multiple dimensions engaged at once (Nursey-Bray & Palmer, 2018).

Nursey-Bray (2022) states that the concept of vulnerability advances victimhood, while the romanticization of resilience erases Indigenous agency and capacity to fully participate in policymaking and decisions. While calls for "co-construction" abound, concerns exist when utilizing "co-construction" about power, conflation of ideas, and loss of Indigenous sovereignty. Indigenous people find these frameworks (of vulnerability and resilience) and tools/metrics not relevant (Singh et al., 2021). A single framework for Indigenous adaptation is hard to achieve, given the context-based solution space where Indigenous people are situated (Caretta, 2021). Further, adaptation science is not constructed by Indigenous people; however, concepts of climate change and adaptation correlate to an Indigenous enduring pattern of "survival and existence that has been part of Indigenous lives and world views always" (Nursey-Bray, 2023: 7).

While many of the chapter authors have adopted a decolonial methodological framework to inform their chapters, this lens is amplified in this chapter by considerations of power. Power asymmetries have been identified as critical barriers to transformative change toward sustainability (Folke, 2016). Power can be defined as all visible, hidden, and invisible dynamics that influence decision-making processes directly or indirectly (McIlwain et al., 2023), influencing decision-making processes and the capacity of different actors to access resources and influence goals, processes, and outcomes of decision-making (Morrison et al., 2019). Power can be assessed to determine who participants, winners, and losers are in decision-making; who controls and sets the agenda of decision-making; and who sets the ideological narrative or "description" of issues (Adapted from Lukes, 2005 and Dahl, 1957).

Indigenous Actions for the Future

The important contribution of this book are the detailed descriptions of the actions Indigenous people take to counter climate change and the increasing cyclones, floods, sea level rise, and landslides they are experiencing. Decolonizing disasters is an important activity to address the inordinate severity of the impacts of these events on Indigenous peoples. Land and community-based values and sustainable practices offer important lessons for addressing climate change in the future. Most important is the dismantling of systems of oppression and the empowerment of Indigenous people in the decision-making processes. Several utterly important steps resonate:

Ending Exclusion

To enhance Indigenous peoples' ability to plan for and respond to climate disasters their voices must be included. This is the crucial first step required. In addition, the physical exclusion of Indigenous peoples from shelters during times of extreme events including flooding must be remedied. Chapters concerning the Munda people document the exclusion of Indigenous people from both the decision-making and the shelters during disasters and an actual "erasure" of their names from the local list of flood-affected people. The Marma of the Chittagong Hill Tracts (CHTs) were excluded from government assistance to rebuild their homes after the 2023 flood; neighboring districts were not so excluded. Indigenous people experience exclusion and inequality outside of disaster planning response. For instance, Indigenous elders of the Marma CHTs are excluded from receiving old-age assistance that other Bangladesh elders receive. To address power deficits and exclusions in disaster situations, structural inequality and exclusion need to be addressed. This is very evident in relation to land.

Recognize Indigenous Land Rights

Many Indigenous peoples are without land rights and have no means of protecting their ancestral lands and the culture that is embodied in and with the land. The circumstances of the Manipuri and the Marma highlight the devastating impact of this lack of recognition. As a result, further incursion on Indigenous land in Bangladesh occurs in the form of land grabbing.

Education Paradigm Shift

Education is a key theme that reappears in the chapters as a solution. Retaining Indigenous culture is required through education, but challenging in government schools where Muslim practices exist in solitude. For the Manipuri, conversion to Christianity is displacing the education of the Manipuri youth of their Indigenous land-based culture and practices. In some contexts, Indigenous women require more education as many Munda women were found to marry too early and without education.

Educating Indigenous communities about their traditions and land-based practices is an important action for retaining Indigenous culture, but also educating non-Indigenous people is equally important. The important adaptations and "relational" learning about living peacefully with nature, and nature's solutions through Indigenous land-based practices to climate change (such as the Manipuri house designing practices) could bring about a paradigm shift solution to climate change. However, adopting Indigenous solutions in a piecemeal manner without Indigenous axiology risks expanding the list of good climate "adaptations" and "actions" to the otherwise-growing list without solving the bigger climate change issue. Perhaps one of the most important learnings is that of empowering Laitu Khyeng Indigenous youth in the CHTs to counter historical injustices, land dispossession, and socio-political marginalization that have disempowered Indigenous people and youth in planning for and responding to climate impact disasters.

Conclusion and Future Work

This book documents oral histories and land-based traditions of Indigenous Bangladesh people and demonstrates their importance for adaptation to climate change now and in the future. The stories and worldviews highlighted in each chapter should continue to be disseminated through traditional media, conferences, and publications but also through storytelling by readers. Portions of each chapter or specific land-based practices (such as building houses in the Manipuri culture) should not be dissected from the holistic cultural and land-based teachings from which they are created. In honor of the relational ontology and epistemology from which these stories were attained, their meanings and learnings should be authentically continued.

Further communication and education surrounding the contents of this book should occur in the widest possible form to the widest possible audience. However, where possible, the participants in the research themselves and the authors should be the source of this knowledge giving power and acknowledgment to their voices. More work documenting the oral histories and traditions of the Indigenous Bangladesh people, and more analysis of laws, power, and justice for these people, is needed.

The Indigenous knowledge recounted in this book evidences the sustainable management of natural resources, especially land, of the Indigenous people in Bangladesh. This knowledge advances autonomous adaptation to climate variability and change. This knowledge provides co-benefits of sustainable food production and biodiversity and ecosystem resilience. However, care to prevent constraints on this knowledge and practice is warranted. The scale of climate change is surpassing the protection offered by some of these practices. As floods and coastal inundation exceed historical records and Indigenous memory, the tipping point might be reached where housing and community infrastructure is lost and the capacity to rebuild and replace is adequately exacerbating migration and displacement. A wicked spiral of loss of Indigenous knowledge might then be triggered where important Indigenous knowledge and land-based practices are further displaced across generations and perhaps ultimately lost. Advancing the important work of this book's authors and researchers, and recognizing the inherently valuable Indigenous land-based practices in decision-making for climate impact and disaster response into the future, is essential in addressing climate change.

References

Adnan, S. (2004). *Migration land alienation and ethnic conflict: Causes of poverty in the 2011 to 2016*. Government of the People's Republic of Bangladesh.

Adnan, S. (2010). Indigenous knowledge and sustainable development: Case studies from Bangladesh. *Journal of Environmental Management*, 91(4), 873–880. https://doi.org/10.1016/j.jenvman.2009.10.014

Adnan S., & Dastidar R. (2011). *Mechamisms of land alienation of the indigenous peoples of the Chittagong Hill Tracts, report of study undertaken for the Chittagong Hill Tracts commission*. IWGIA.

ADRC (2024). Hot-compression molding temperature of bamboo-based fiber composites. *Applied Sciences*, 14(14), 6080.

Ahamed, F. U. (2001). *Ethnicity and environment:'Tribal'culture and the state in Bangladesh*. University of London, University College London.

Ahammad, R. (2019). Recent trends in forest and livelihood relationships of rural communities in the Chittagong Hill Tracts region, Bangladesh. Doctoral dissertation, Charles Darwin University, Australia.

Ahammad, R., Stacey, N., Eddy, I.M., Tomscha, S.A., Sunderland, T.C. (2019). Recent trends of forest cover change and ecosystem services in eastern upland region of Bangladesh. *Science of the Total Environment*, 647, 379–389.

Ahmed, B., & Alam, M. (2021). Cyclone shelters and disaster risk reduction in coastal Bangladesh: A critical analysis. *Disaster Prevention and Management*, 30(4), 377–390. https://doi.org/10.1108/DPM-07-2020-0240

Ahmed, S., & Eklund, E. (2021, January 4). Climate change impacts in Coastal Bangladesh: Migration, gender and environmental injustice. *Asian Affairs*, 52(1), 155–174. https://doi.org/10.1080/03068374.2021.1880213

Ahmed, B., Kelman, I., Fehr, H.K., Saha, M. (2016). Community resilience to cyclone disasters in coastal Bangladesh. *Sustainability*, 8, 805.

Ahmmed, M. F., & Singh, L. (2007). State of the rural Manipuris in Bangladesh. http://www.ecdo-bd.org/Resources/Rural_Manipuris.pdf

Akinola, O. V., Adegoke, J., & Ologunorisa, T. E. (2019). Assessment of social vulnerability to wildfire in Missouri, United States of America. *Journal of Sustainable Development*, 12(4), 76. https://doi.org/10.5539/jsd.v12n4p76

Akter, S., Howladar, M. F., Ahmed, Z., & Chowdhury, T. R. (2019). The rainfall and discharge trends of Surma River area in North-eastern part of Bangladesh: An approach for understanding the impacts of climatic change. *Environmental Systems Research*, 8(1), 1–12.

Alam, Z. (2023). Damage and Needs Assessment CHT Monsoon Rain and Flash Flood Final Report, https://sheltercluster.org/bangladesh-flash-flood-and-monsoon-rain-2023/documents/damage-and-needs-assessmentcht-monsoon-rain

Alam, G. M. M., Alam, K., & Mushtaq, S. (2022). Climate change and food security in Bangladesh: A comprehensive review. *Environmental Science & Policy*, 133, 144–154. https://doi.org/10.1016/j.envsci.2022.02.016

Alestra, C., Cette, G., Chouard, V., & Lecat, R. (2023). How can technology significantly contribute to climate change mitigation? *Applied Economics*, 56(41), 4925–4937. https://doi.org/10.1080/00036846.2023.2227416

Ali, T., Paton, D., Buergelt, P. T., Smith, J. A., Jehan, N., & Siddique, A. (2021). Integrating Indigenous perspectives and community-based disaster risk reduction: A pathway for sustainable Indigenous development in Northern Pakistan. *International Journal of Disaster Risk Reduction*, 59, 102263.

Álvarez, L., & Coolsaet, B. (2020). Decolonizing environmental justice studies: A Latin American perspective. *Capitalism Nature Socialism*, 31(2), 50–69.

Archibald, J. A. (2008). *Indigenous storywork: Educating the heart, mind, body, and spirit*. UBC Press.

Ashrafuzzaman, M. (2014). *The Tragedy of the Chittagong Hill Tracts in Bangladesh: Land Rights of Indigenous People*. Lund: Lund University.

Ashrafuzzaman, M., Gomes, C., & Guerra, J. (2022, August 19). Climate justice for the southwestern coastal region of Bangladesh. *Frontiers in Climate*, 4. https://doi.org/10.3389/fclim.2022.881709

Asia Disaster Reduction Center (2023). https://www.adrc.asia/view_disaster_en.php?NationCode=50&Lang=en&Key=2632

Assaduzzaman, M., Filatova, T., Lovett, J. C., & Coenen, F. H. J. M. (2023, February 17). Gender-ethnicity intersectionality in climate change adaptation in the coastal areas of Bangladesh. *Sustainability*, 15(4), 3744. https://doi.org/10.3390/su15043744

Asugeni, R., Redman-MacLaren, M., Asugeni, J., Esau, T., Timothy, F., Massey, P., & MacLaren, D. (2017). A community builds a "bridge": An example of community-led adaptation to sea-level rise in East Kwaio, Solomon Islands. *Climate and Development*, 11(1), 91–96.

Austin, D. (2004). Partnerships, not projects! Improving the environment through collaborative research and action. *Human Organization*, 63(4), 419–430.

Axelsson, P., & Sköld, P. (2006). Indigenous populations and vulnerability. Characterizing vulnerability in a Sami context. In *Annales de démographie historique* (No. 1, pp. 115–132). Cairn/Softwin.

Ayeb-Karlsson. (2020). 'When the Disaster Strikes: Gendered (Im)mobility in Bangladesh'. *Climate Risk Management*, 29, 100237.

Baird, I. G. (2008). Local knowledge and natural resource management: Perspectives from Southeast Asia. *Asian Journal of Social Science*, 36(3), 342–356. https://doi.org/10.1163/156853108X331818

Banerjee, A. K. (2000). Devolving forest management in Asia-Pacific countries. In T. Enters, P. B. Durst, & M. Victor (Eds.), *Decentralization and devolution of forest management in Asia and the Pacific*. RECOFTC Report 18, RAP Publication 2000/1, Bangkok.

Bang, M., Curley, L., Kessel, A., Marin, A., Suzukovich III, E. S., & Strack, G. (2014). Muskrat theories, tobacco in the streets, and living Chicago as Indigenous land. *Environmental Education Research*, 20(1), 37–55.

Barkat, A., Halim, S., Osman, A., Hossain, I., & Ahsan, M. (2010). *Status and dynamics of land rights, land use and population in Chittagong Hill Tracts of Bangladesh*. Human Development Research Centre.

Barua, B. (2004). Western education and modernization in a Buddhist village of Bangladesh: A case of the Barua community. Unpublished Ph.D dissertation, Department of Sociology and Equity St.

Barua, D., & Ando, M. (2002). *Syncretism in Bangladeshi Buddhism*. Chubu Nihon Kyoiku Bunkai.

Battiste, M. (2014). Decolonizing education: Nourishing the learning spirit. *Alberta Journal of Educational Research*, *60*(3), 615–618.
Battiste, M. (2017). *Decolonizing education: Nourishing the learning spirit*. UBC Press.
BBS. (2022). Bangladesh Bureau of Statistics, Government of the People's Republic of Bangladesh.
Bdnews24.com (2024). In Bandarban's Chimbuk Hill, a community quietly struggles through a water crisis https://bdnews24.com/bangladesh/7o7mkc8eyr
Berkes, F. (2018). *Sacred ecology* (4th ed.). Routledge.
Bhambra, G. K. (2020). Colonial global economy: Towards a theoretical reorientation of political economy. *Review of International Political Economy*, 28(2), 307–322.
Bhattacharjee, J., Bhattacharjee, D., Paul, T., Kumar, A., & Chowdhury, S. (2015). Diversity of mushrooms in Indo-Bangladesh region of north-East India. *Journal of the Andaman Science Association*, 19(1), 75–82.
Bowra, A., Saeed, G., Gorodensky, A., & Kohler, J. C. (2022). An exploration of anti-corruption and health in international organizations. *Plos one*, 17(8), e0269203.
Bronen, R. (2011). Climate-induced community relocations: Creating an adaptive governance framework based in human rights doctrine. *N. Y. U. Review of Law & Social Change*, 35, 357.
Brown, et al. (2016). Enhancing Healthcare Equity with Indigenous Populations: Evidence-Based Strategies from an Ethnographic Study. *BMC Health Services Research*, 16, 1–17.
Bryman, A. (2016). *Social research methods*. Oxford University Press.
Caretta, L., Oh, S. H., Fakhrul, T., Lee, D. K., Lee, B. H., Kim, S. K., … Beach, G. S. (2021). Relativistic kinematics of a magnetic soliton. *Science*, 370(6523), 1438–1442.
Caretta, M., & Chalmers, J. (2017). The transformation of academic knowledges: Understanding the relationship between decolonising and indigenous research meth odologies. *Socialist Studies*, 12(1), 97–116. https://doi.org/10.18740/S4GH0C DOI: 10.1080/17565529.2021.1993627
Chakma, B. (2010a). The post-colonial state and minorities: Ethnocide in the Chittagong Hill Tracts, Bangladesh. *Commonwealth & Comparative Politics*, 48, 281–300.
Chakma, K. (2010b). Indigenous rights and research ethics: Perspectives from Bangladesh. *Journal of Indigenous Studies*, 15(2), 43–56.
Chakma, J. (2022). A case Study on the Popular Rituals of Marma Buddhists in Bangladesh. *Social Science Review* [The Dhaka University Studies, Part-D], 39(1), June 2022 https://doi.org/10.3329/ssr.v39i1.64920
Chakma, N., & Maitrot, M. (2016). How ethnic minorities became poor and stay poor in Bangladesh: A qualitative enquiry. EEP/Shiree, July.
Chakma, S., & Hokugo, A. (2020). Evacuation behavior: Why do some people never evacuate to a cyclone shelter during an emergency? A case study of coastal Bangladesh. *Journal of Disaster Research*, 15(4), 481–489.
Chakma, U., & Sultana, S. (2021). Colonial governmentality and Bangladeshis in the anthropocene: Loss of language, land, knowledge, and identity of the Chakma in the ecology of the Chittagong Hill tracts in Bangladesh. *Ethnicities*, 24(4), 560–580.
Chakma, U., & Sultana, S. (2023). Colonial governmentality and Bangladeshis in the anthropocene: Loss of language, land, knowledge, and identity of the Chakma in the ecology of the Chittagong Hill tracts in Bangladesh. *Ethnicities*, 14687968231219521.
Chakraborty, S. K., & Choudhury, D. (2023). Drawing on Past Practices to Secure the Future: Innovative applications of traditional lac rearing in the fallows of Karbi farmers in Assam, northeast India. In *Farmer Innovations and Best Practices by Shifting Cultivators in Asia-Pacific* (pp. 690–712). GB: CABI.
Chapola, J. (2022). A Racialized Settler Woman's Transformative Journey In Canada: Building Relational Accountabilities (Doctoral dissertation, University of Saskatchewan).

Chen, J. J., & Mueller, V. (2018). Coastal Climate Change, Soil Salinity and Human Migration in Bangladesh. *Nature Climate Change*, 8, 981–985.

Chilisa, B. (2020). Indigenous research methodologies. *Handbook of Research Methods in Health Social Sciences*, 49–64. https://doi.org/10.1007/978-981-10-2779-6_49-1

Chowdhury, M. R., Mollah, M. B., & Hossain, M. M. (2021). Coastal community resilience to cyclones in Bangladesh: A case study of cyclone Aila. *Journal of Environmental Management*, 289, 112537. https://doi.org/10.1016/j.jenvman.2021.112537

Climate Change Report (2020). www.manusherjonno.org/wp-content/uploads/2021/01/Climate-Change-Report-2020.pdf. Retrieved on May 15, 2023.

Corntassel, J., & Hardbarger, T. (2019). Educate to perpetuate: Land-based pedagogies and community resurgence. *International Review of Education*, 65, 87–116.

Coulthard, G. S. (2014). *Red skin, white masks: Rejecting the colonial politics of recognition*. Minneapolis: Minnesota.

Crosweller, M., & Tschakert, P. (2020). Disaster management leadership and policy making: A critical examination of communitarian and individualistic understandings of resilience and vulnerability. *Climate Policy*, 21(2), 203–221.

Dahl, R. A. 1957. The concept of power. *Behavioral Science*, 2(3), 201–215. https://doi.org/10.1002/bs.3830020303

Das, S. (2011). Indigenous people's access to land in Northern-belt of Bangladesh: A study of the Santal community (Master's thesis, Universitetet i Tromsø).

Datta, R., (2015). A relational theoretical framework and meanings of land, nature, and sustainability for research with Indigenous communities. *Local Environment: The International Journal of Justice and Sustainability*, 20, 102–113.

Datta, R. (2017). Traditional storytelling: An effective indigenous research methodology and its implications for environmental research. *AlterNative: An International Journal of Indigenous Peoples*, 14, 35–44. BMC Health Serv. Res, pp. 16, 1–17.

Datta, R. (2018a). Decolonizing both researcher and research and its effectiveness in Indigenous research. *Research Ethics*, 14(2), 1–24.

Datta, R. (2018b). Traditional storytelling: An effective Indigenous research methodology and its implications for environmental research. *AlterNative: An International Journal of Indigenous Peoples*, 14(1), 35–44.

Datta, R. (2019). *Land-water management and sustainability in Bangladesh: Indigenous practices in the Chittagong Hill Tracts*. Routledge.

Datta, R. (2023). Land-based environmental sustainability: A learning journey from Indigenist researcher. *Polar Geography*, 46(1), 3–17.

Datta, R. (2024a). *Decolonization in practice: A cross-cultural perspective*. Canada Scholar Press.

Datta, R. (2024b). Relationality in indigenous climate change education research: A learning Journey from indigenous communities in Bangladesh. *Australian Journal of Environmental Education*, 40(2), 128–142. doi:10.1017/aee.2024.13

Datta, R., & Chapola, J. (2018). Indigenous and Western Environmental Resource Management: A Learning Experience with the Laitu Khyeng Indigenous Community in the Chittagong Hill Tracts (CHT), Bangladesh. *Native American and Indigenous Studies*, 4(2), 52–80.

Datta, R., Chapola, J., Owen, K., Hurlbert, M., & Foggin, A. (2024). Indigenous land-based practices for climate crisis adaptions. *Explore*, 20(6), 103042.

Datta, R., Chapola, J., Waucaush-Warn, J., Subroto, S., & Hurlbert, M. (2024). Decolonizing meanings climate crisis and land-based adaptions: From Indigenous women's perspectives in Western Canada. Women's Studies International Forum.

Datta, R., & Kairy, B. (2024). Decolonizing climate change adaptations from indigenous perspectives: Learning reflections from Munda indigenous communities, Coastal Areas in Bangladesh. *Sustainability*, 16(2), 769. https://doi.org/10.3390/su16020769

Datta, R., & Marion, W. (2021). Ongoing Colonization and Indigenous Environmental Heritage Rights: A Learning Experience with Cree First Nation Communities, Saskatchewan, Canada. *Heritage*, 4, 1388–1399. 10.3390/heritage4030076

Demos, T. J. (2020). *Beyond the world's end: arts of living at the crossing*. Duke University.

Denzin, N. K., Lincoln, Y. S., & Smith, L. T. (Eds.). (2008). *Handbook of critical and indigenous methodologies*. Sage.

Dewan, C. (2021). Misreading the Bengal delta: Climate change, development, and livelihoods directions for emerging scholarship. *Environment and Planning E: Nature and directions. Watershed Ecology and the Environment*, 1, 42–56.

Dhaka Tribune (2023). Empowering Kheyang community: Native 'Hyow' script keyboard launched www.dhakatribune.com/bangladesh/nation/326134/empowering-kheyang-community-native-hyow-script Retrieved on May 15, 2023.

DoE (Department of Environment). (2015). Fifth National Report to the Convention on Biological Diversity. Department of Environment, Ministry of Environment and Forests, Government of the People's Republic Bangladesh, Dhaka.

Enarson, E., & Chakrabarti, P. G. D. (2009, December 10). *Women, Gender and Disaster*. SAGE Publications India.

Ermine, W. (1995). Aboriginal epistemology. In M. Battiste & J. Barman (Eds.), *First Nations education in Canada: The circle unfolds* (pp. 101–112). UBC Press.

Fardous, S. (2011). Perception of climate change in Kaptai National Park. Rural Livelihoods and Protected Landscape: Co-Management in the Wetlands and Forests of Bangladesh, 186–204.

Feroz, S.M., Alam, M.R., Das, P., & Al Mamun, A. (2014). Community ecology and spatial distribution of trees in a tropical wet evergreen forest in Kaptai national park in Chittagong Hill Tracts, Bangladesh. *Journal of Forest Research*, 25(2), 311–318.

Folke, C. (2016). Resilience (Republished). *Ecology and Society*, 21(4). https://doi.org10.5751/ES-09088-210444

Gaillard, J.C. (2012). The Climate Gap, *Climate and Development Publications*, 4, 261–264.

Galway, L. P., Esquega, E., & Jones-Casey, K. (2022). "Land is everything, land is us": Exploring the connections between climate change, land, and health in Fort William First Nation. *Social Science & Medicine*, 294, 114700.

Garai, J., Ku, H. B., & Zhan, Y. (2022). Climate change and cultural responses of indigenous people: A case from Bangladesh. *Current Research in Environmental Sustainability*, 4, 100130.

Garai, S., Bhattacharjee, C., Sarkar, S., Moulick, D., Dey, S., Jana, S., & Hossain, A. (2022). Microplastics in the soil–water–food nexus: Inclusive insight into global research findings. *Science of the Total Environment*, 946, 173891.

Garai, S., Majumder, A., & Sinha, S. (2022). *Community-based participatory research and Indigenous knowledge systems: Case studies from South Asia*. Springer.

Goldman, M. J., Turner, M. D., & Daly, M. (2018). A critical political ecology of human dimensions of climate change: Epistemology, ontology, and ethics. *WIREs Climate Change*, 9(4), e526. https://doi.org/10.1002/wcc.526

Gonzalez, C.G. 2013. Racial capitalism, climate change, and Ecocide. *Wisconsin International Law Journal*. 479–520. Available at: https://doi.org/10.59015/wilj.VCPQ4704

Gonzalez, C. G. (2024). Racial capitalism, climate change, and ecocide. *Climate Change, and Ecocide* (April 21, 2024), 41.

Gordon, H. S. J. & Datta, R. (2022). Indigenous communities defining and utilising self-determination as an individual and collective capability. *Journal of Human Development and Capabilities*, 23(2), 182–205. https://doi.org/10.1080/19452829.2021.1966613

Greater Wellington (2021). Integrating native planting and flood protection: an operational guide for Greater Wellington. https://www.gw.govt.nz/assets/CAMA.FLDP.2020.J001105-River-berm-planting-guide-1.7.pdf Retrieved on May 15, 2023.

Greene, E. A. (2021). Learning the Land: An Intergenerational Study of Inuit Knowing (Doctoral Dissertation, University of British Columbia).

Gunter, B. G., A. Rahman, and A. F. M. A. Rahman. (2008). *How Vulnerable are Bangladesh's Indigenous People to Climate Change?* Bangladesh Development Research Center.

Haque, M. N., Baroi, A., Gomes, J., Toppo, A., Das, R. S., & Hossain, M. K. (2021a). Effects of climate change and analyzing the Indigenous practices for adaptation to climate change impacts in Chittagong Hill tracts (CHT) of Bangladesh. *Asian Journal of Science and Technology*, 12(10), 11856–11864.

Haque, S. E., Gazi-Khan, L., & Rashid, H. (2021b). Climate change mitigation and adaptation strategies, the environment, and impacts of the COVID-19 pandemic: A review of the literature. *Role of Green Chemistry in Ecosystem Restoration to Achieve Environmental Sustainability*, 61–72.

Haque, U., Hashizume, M., Kolivras, K. N., Overgaard, H. J., Das, B., & Yamamoto, T. (2021c). Reduced death rates from cyclones in Bangladesh: What more needs to be done? *Bulletin of the World Health Organization*, 100(1), 81–90. https://doi.org/10.2471/BLT.21.286641

Haque, A., & Mondal, M. (2021). Indigenous knowledge-based adaptation to climate change in the coastal regions of Bangladesh. *Environmental Development*, 38, 100619. https://doi.org/10.1016/j.envdev.2021.100619

Haque, M. N., Toppo, A., Das, R. S., & Ratan, M. (2022). Impact of agroecology in the agricultural activities of Chittagong Hill Tract (CHT) for improving local food production and livelihoods in Bangladesh. *Journal of Bangladesh Agriculture*, 12(1), 117–129.

Hart, M. A. (2010). Indigenous worldviews, knowledge, and research: The development of an Indigenous research paradigm. *Journal of Indigenous Voices in Social Work*, 1(1), 1–16.

Hiwasaki, L., Luna, E., Syamsidik, & Shaw, R. (2014). *Local and indigenous knowledge for community resilience: Hydro-meteorological disaster risk reduction and climate change adaptation in coastal and small island communities*. UNESCO.

Hoque, M. Z., Cui, S., Lilai, X., Islam, I., Ali, G., & Tang, J. (2019). Resilience of coastal communities to climate change in Bangladesh: Research gaps and future directions. *Watershed Ecology and the Environment*, 1, 42–56. https://doi.org/10.1016/j.wsee.2019.10.001

Hossain, M.K. (1998). Role of Plantation Forestry in the Rehabilitation of Degraded and Secondary Hill Forests of Bangladesh. In: Proceedings of the IUFRO Inter-Divisional Seoul Conference-Forest Ecosystem and Land use in Mountain Areas, 12–17 October, 1998, Seoul, Korea. pp. 243–250.

Hossain, M.K., Alam, M.K., Miah, M.D. (2008). Forest restoration and rehabilitation in Bangladesh. In: D.K. Lee (Ed.), *Keep Asia Green, IUFRO World Series Volume 20-III "South Asia"*. IUFRO, pp. 21–66.

Hossain, M. S., & Deb, U. K. (2021). Community-based adaptation to climate change in Bangladesh: Achievements and challenges. *Journal of Environmental Management*, 299, 113686. https://doi.org/10.1016/j.jenvman.2021.113686

Hossain, M. S., & Sakai, T. (2021). Vulnerability and adaptation to climate change in coastal Bangladesh. *Sustainability*, 13(7), 3765. https://doi.org/10.3390/su13073765

Hossain, M., Saha, C., Rubaiot Abdullah, S. M., Saha, S., & Siddique, M. R. H. (2016). Allometric biomass, nutrient and carbon stock models for Kandelia candel of the Sundarbans, Bangladesh. *Trees*, 30, 709–717.

Hunsberger, C., Corbera, E., Borras Jr, S. M., Franco, J. C., Woods, K., Work, C., de la Rosa, R., Eang, V., Herre, R., & Kham, S. S. (2017). Climate change mitigation, land grabbing and conflict: towards a landscape-based and collaborative action research agenda. *Canadian Journal of Development Studies/Revue canadienne d'études du développement*, 38(3), 305–324.

Iftekhar, M.S., & Hoque, A.K.F. (2005). Causes of forest encroachment: An analysis of Bangladesh. *GeoJournal*, 62(1), 95–106. https://doi.org/10.1007/s10708-005-7917-z

Imperiale, A. J., & Vanclay, F. (2024). Understanding the social dimensions of resilience: The role of the social sciences in disaster risk reduction, climate action, and sustainable development. *Sustainable Development*, 32(2), 1371–1375.

IPCC (Intergovernmental Panel on Climate Change). (2007). Contribution of Working Group II to the Fourth Assessment Report of the Intergovernmental Panel on Climate Change: Summary for Policymakers. Intergovernmental Panel on Climate Change: Brussels, Belgium.

IPCC. (2014). Annex II: Glossary in: Climate change 2014: Impacts, adaptation, and vulnerability. In *Contribution of working group II to the fifth assessment report of the intergovernmental panel on climate change* (pp. 1757–1776). Cambridge University Press.

IPCC (2023). Intergovernmental Panel on Climate Change (IPCC). Summary for Policymakers. In *Climate change 2020: The physical science basis. Contribution of working group I to the sixth assessment report of the intergovernmental panel on climate change* (pp. 3–32). Cambridge University Press.

Islam, M. R., Ali, A., & Khan, H. A. (2022). Integrating Indigenous knowledge into flood risk management: Insights from Bangladesh. *International Journal of Disaster Risk Reduction*, 76, 103005. https://doi.org/10.1016/j.ijdrr.2022.103005

Islam, S. D. U., Bhuiyan, M. A., & Ramanathan, A. (2015). Climate Change Impacts and Vulnerability Assessment in Coastal Region of Bangladesh: A Case Study on Shyamnagar Upazila of Satkhira District. *Journal of Climate Change*, 1(1,2), 37–45. https://doi.org/10.3233/jcc-150003

Islam, M. S., Hasan, G. J., & Chowdhury, M. A. I. (2006). Destroying hills in the northeastern part of Bangladesh: A qualitative assessment of extent of the problem and its probable impact. *International Journal of Environmental Science & Technology*, 2, 301–308.

Islam, A.R., Hasan, M.M., Islam, M.T., & Tanaka, N. (2022). Ethnobotanical study of plants used by the Munda ethnic group living around the Sundarbans, the world's largest mangrove forest in southwestern Bangladesh. *Journal of Ethnopharmacology*, 285, 114853.

Islam, K.K., & Hyakumura, K. (2021). The potential perils of Sal forests land grabbing in Bangladesh: an analysis of economic, social and ecological perspectives. *Environment Development and Sustainability* 23(1), 1–23. DOI:10.1007/s10668-021-01301-7

Islam, M. R., & Shamsuddoha, M. (2018). Socioeconomic implications of climate change for Bangladesh. *Asian Development Review*, 35(2), 94–118. https://doi.org/10.1162/adev_a_00111

Johnson, D. E., Parsons, M., & Fisher, K. (2022). *Indigenous climate change adaptation: New justice and community engagement*. Routledge.

Kapoor, K. (2008). *The Postcolonial Politics of Development*, Routledge.

Karistie, J. F., Ridho, A., Pangestika, F. A., & Bisri, M. B. F. (2023). Women participation in disaster risk management practices in Indonesia: A systematic review. Easychair.org.

Karlsson, S. (2020). Transforming the climate change-gender nexus?: Revisiting gender mainstreaming in the context of climate change and disaster risk reduction: experiences from the Swedish Embassy in Bangkok.

Khan, A. M. (1999). *The Maghs: A Buddhist Community in Bangladesh*. University Press Limited.

Khan, Z.H. (2015). Study on water supply system at Rangamati Municipal Area. *Journal of Environmental Engineering*, 3(1), 15–22.

Khan, A., Fux, B., Su, C., Dubey, J. P., Dardé, M. L., Ajioka, J. W., … Sibley, L. D. (2007). Recent transcontinental sweep of Toxoplasma gondii driven by a single monomorphic chromosome. *Proceedings of the National Academy of Sciences*, 104(37), 14872–14877.

Khisa, S. K., & Mohiuddin, M. (2006). Shrinking jum and changing livelihoods in the Chittagong Hill Tracts of Bangladesh. *Shifting Cultivation, Livelihood and Food Security*, 41.

Kimmerer, R. (2013). *Braiding sweetgrass: Indigenous wisdom, scientific knowledge and the teachings of plants*. Milkweed editions.

Kirkby, P., Williams, C., & Huq, S. (2015). A brief overview of Community-Based Adaptation. International Centre for Climate Change and Development: Briefing paper.

Kovach, M. (2009). *Indigenous Methodologies: Characteristics, Conversations, and Contexts*. University of Toronto Press.

Kovach, M. (2010a). *Indigenous Methodologies: Characteristics, Conversations, and Contexts*. University of Toronto Press.

Kovach, M. (2010b). Conversational method in Indigenous research. *First Peoples Child and Family Review*, 5, 40–48.

Kovach, M. (2016). Doing indigenous methodologies. *The SAGE Handbook of Qualitative Research*, 383–406.

Kovach, M. (2021). *Indigenous methodologies: Characteristics, conversations, and contexts*. University of Toronto Press.

Leonard, K. (2021). WAMPUM Adaptation framework: Eastern coastal Tribal Nations and sea level rise impacts on water security. *Climate and Development*, 1–10. https://doi.org/10.1080/17565529.2020. 1862739

Lewis, J. (2003). Design issues. In J. Ritchie & J. Lewis (Eds.), Qualitative research practice SAGE publications.

Little Bear, L. (2013). Indigenous knowledge and the Western world: A critical analysis. In N. J. (Ed.), *Knowledge, learning, and Indigenous knowledge systems* (pp. 34–47). Oxford University Press.

Lukes, S. (2005). *Power: A radical view* (2nd ed.). Palgrave Macmillan.

Mahmood, H., Siddique, M. R. H., Rubaiot Abdullah, S. M., Costello, L., Matieu, H., Iqbal, M. Z., & Akhter, M. (2019). Which option best estimates the above-ground biomass of mangroves of Bangladesh: Pantropical or site-and species-specific models? *Wetlands Ecology and Management*, 27, 553–569.

Mahmud, I. (2024, 31 January). Over 4,000 illegal brick kilns. Prothom-alo. https://en.prothomalo.com/environment/vks7nmkf00

Mahmud et al. (2017). The effects of Teak monoculture on forest soils: A case study in Bangladesh. *Journal of Forest Research*, 29, 1111–1120 (2018). https://doi.org/10.1007/s11676-017-0515-3

Maldonado, J. K., Shearer, C., Bronen, R., Peterson, K., & Lazrus, H. (2014). The impact of climate change on tribal communities in the US: Displacement, relocation, and human rights. *Climate change and indigenous peoples in the United States: Impacts, experiences and actions*, pp. 93–106.

Mallick, B. (2019). The nexus between socio-ecological system, livelihood resilience, and migration decisions: Empirical evidence from Bangladesh. *Sustainability*, 11(12), 3332.

Markham, F., Smith, D. & Morphy, F. (2020). Indigenous Australians and the COVID-19 Crisis: Perspectives on Public Policy. Topical Issue no. 1/2020, Centre for Aboriginal Economic Policy Research, Australian National University, Canberra.

Marma, A. (2009). The Livelihood of Marma Community in Bangladesh. Bukkyo Bunka Kenkyo Sho, Vol. 48, pp. 17–22. Kyoto: Ryukoku University.

Mason, S., & Rigg, J. (2019). *Revisiting Indigenous knowledge systems and community engagement: New perspectives and case studies*. Routledge.

Mathbor, G. M. (2007). Enhancement of community preparedness for natural disasters. *International Social Work*. https://doi.org/10.1177/0020872807076049

McGregor, D. (2014). Traditional knowledge and water governance: the ethic of responsibility. *AlterNative*, 10(5), 493–507.

McGregor, D. (2018). From "Decolonized" to Reconciled Research: Moving from Decolonizing Research to Reconciliation Research in Canada. *AlterNative: An International Journal of Indigenous Peoples*, 14(3), 218–226. https://doi.org/10.1177/1177180118792332

McIlwain, L., Holzer, J., Baird, J., & Baldwin, C. (2023). Power Research in Adaptive Water Governance and beyond: A Review. *Ecology and Society*, 28(2). https://doi.org/10.5751/es-14072-280222

McVittie, J. et al. (2019). Relationality and Decolonization in Children and Youth Garden Spaces. *Australian Journal of Environmental Education*, 35, 93–109. [CrossRef].

Mey, W. (1984). (Ed). Genocide in the Chittagong Hill Tracts, Bangladesh. Document No. 51, Copenhagen: International Workgroup for Indigenous Affairs (IWGIA).

Mignolo, W. D., & Walsh, C. E. (2018). *On decoloniality: Concepts, analytics, praxis*. Duke University Press.

Ministry of Health and Family Welfare. (2016). Tribal/Ethnic Health Population and

Mintenbeck, A. Alegría, M. Craig, S. Langsdorf, S. Löschke, V. Möller, A. Okem, B. Rama (eds.)]. (2023). Cambridge University Press, Cambridge, UK and New York, NY, USA, pp. 1457–1579, doi:10.1017/9781009325844.012

MoFA (2018). Climate Change Profile: Bangladesh. Ministry of Foreign Affairs, Government of the People's Republic of Bangladesh.

Mohiuddin, M. (2009). Studies on Traditional Knowledge of Plant Uses and Their Conservation Prospects by Upland Communities in Bandarban Hill District, Bangladesh, PhD Thesis submitted to the University of Chittagong, Bangladesh.

Mohiuddin, S., Saadat, S., & Nisar, N. (2012). Oral health knowledge and oral hygiene practices among private and public school teachers of Karachi city. *Journal of the Dow University of Health Sciences (JDUHS)*, 9(2), 64–70.

Mohsin, A. (1995). The politics of nationalism: the case of the Chittagong Hill Tracts, Bangladesh, PhD thesis, University of Cambridge, Cambridge.

Mohsin, A. A. (2002). Religion, politics and security: The case of Bangladesh. In Satu P. Limaye, Robert G. Wirsing, Mohan Malik (Eds.), *Religious Radicalism and Security in South Asia*, 467–488. Asia-Pacific Center for Security Studies.

Morrison, B. D., Heath, K., & Greenberg, J. A. (2019). Spatial scale affects novel and disappeared climate change projections in Alaska. *Ecology and Evolution*, 9(21), 12026–12044.

Muhammed, N., Koike, M., & Bitte, R. A.W., (2004). Investment analysis of teak (Tectona grandis): a case study on Sylhet Forests of Bangladesh. *Journal of Forest Planning*, 10(2), 77–86.

Muhammed, N., Koike, M., Haque, F., Chowdhury, M.S.H., & Halim, M.A., (2007). Assessment of Teak (Tectona grandis) timber sale and its associated price influencing factors: a case study on Sylhet forests of Bangladesh. *International Journal of Sustainable Agricultural Technology*, 3(1), 42–48.

Mukul, S. A., Rashid, A. Z. M. M., & Khan, N. A. (2021). Indigenous people and their livelihoods: Learning from the conservation practices in the Chittagong Hill Tracts, Bangladesh. *Forest Policy and Economics*, 126, 102415. https://doi.org/10.1016/j.forpol.2021.102415

Mungazi, D. A. (1986). Application of Memmi's Theory of the Colonizer and the Colonized to the Conflicts in Zimbabwe. *The Journal of Negro Education*, 55(4), 518–534. https://doi.org/10.2307/2294835

Murphy, B. (2016). Testing the Links between Soil Security, Sustainable Land Management Practices and Land Evaluation. Chapter 10. In Field, Damien, Morgan, Cristine L., McBratney, Alex B. (Eds.) *Global Soil Security Symposium*. Progress in Soil Science Series. (Springer Publishers).

Mustajib, S. (2020). Impact of Climate Change on Indigenous Communities in Bangladesh https://www.internationalaffairshouse.org/impact-of-climate-change-on-indigenous-communities-in-bangladesh/

Muthukrishnan, R., & Datta, R. (2023). *Indigenous practice and community-led climate change solutions: The relevance of traditional cosmic knowledge systems*. Taylor & Francis.

Nasher, N. R., Karim, K. R., & Islam, M. Y. (2022). Spatio-temporal variation of cyclone intensity over the coastal region of Bangladesh using 134 years track analysis. *Tropical Cyclone Research and Review*, 11(1), 16–25.

Newing, H. (2010). *Conducting Research in Conservation: Social Science Methods and Practice*. Routledge.

Niamir, L., & Pachauri, S. (2023, March 3). From social and natural vulnerability to human-centered climate resilient coastal cities. *Frontiers in Sustainable Cities*, 5. https://doi.org/10.3389/frsc.2023.1137641

Nightingale, E., & Richmond, C. A. (2022). Building structures of environmental repossession to reclaim land, self-determination and Indigenous wellness. *Health & Place*, 73, 102725.

Nursey-Bray, M. (2023). Building Collaborative Models for Change: Engaging Indigenous People and Country into in Socially just Conservation. In *Rural Development for Sustainable Social-ecological Systems: Putting Communities First* (pp. 209–227). Springer International Publishing.

Nursey-Bray, M., Palmer, R. (2018). Country, climate change adaptation and colonization: insights from an Indigenous adaptation planning process, Australia. *Heliyon*, 4. https://doi.org/10.1080/17565529.2021.1993627

Nursey-Bray, M., Palmer, R., Chischilly, A.M., Rist, P., Yin, L. (2022). *Old Ways for New Days: Indigenous Survival and agency in Climate changed Times*. Springer Briefs in Climate Studies, Springer, Cham, Switzerland. https://doi.org/10.1007/978-3-030-97826-6

Nutrition Plan for the Health, Population and Nutrition Sector Development Program (HPNSDP) P. (Ed.), (2023). *Survival on the Fringe, Adivasis of Bangladesh* (pp. 43–51). Society for

Paul, S. K., & Routray, J. K. (2022). Post-Cyclone Livelihood Strategies and Security Status of Coastal Households in Bangladesh: An Empirical Study. In *Climate Change and Risk in South and Southeast Asia* (pp. 36–58). Routledge.

PCJSS (2023). Population in CHT, The Parbatya Chattagram Jana Samhati Samiti (PCJSS). https://www.pcjss.org/population-in-cht/

Peltier, C. (2018). An application of Two-Eyed Seeing: Indigenous research methods with participants experiencing homelessness. *International Journal of Qualitative Methods*, 17(1), 1–10. https://doi.org/10.1177/1609406918812346

Pender, J.S. (2008). What Is Climate Change? And How It Will Effect Bangladesh. Briefing

Population & Housing Census (2022). Population & Housing Census 2022 Preliminary Report, Bangladesh bureau of Statistics and Information Division, Ministry of Planning, The Government of the Peoples Republic of Bangladesh, https://sid.portal.gov.bd/sites/default/files/files/sid.portal.gov.bd/publications/01ad1ffe_cfef_4811_af97_594b6c64d7c3/PHC_Preliminary_Report_(English)_August_2022.pdf

Rabbani, M.G. et al. (2018). Changing Climate Hazards in the Coast: Risks and Impacts on Satkhira, One of the Most Vulnerable Districts in Bangladesh, Intechopen: Rijeka Croatia.

Rahman, A. E. M. O. (2019). An analysis of language shift and maintenance among Marma community of Bangladesh (Doctoral dissertation, Brac University).

Rahman, M.S., Ahmed, N.F., Ali, M., Abedin, M.M. & Islam, M.S. (2019). Determinants of tobacco cultivation in Bangladesh. *Tobacco Control*, 29(6), 692–694. https://doi.org/10.1136/tobaccocontrol-2019-055167

Rahman, S., & Biswas, A. (2021). Building resilience through community-based adaptation in rural Bangladesh. *International Journal of Climate Change Strategies and Management*, 13(3), 367–382. https://doi.org/10.1108/IJCCSM-04-2020-0038

Rahman, M. S., & Khan, M. A. (2022). Vulnerability and adaptation to climate change in Bangladesh: A comprehensive review. *Environmental Research*, 208, 112581. https://doi.org/10.1016/j.envres.2022.112581

Rahman, M. S., Khan, M. A., & Haque, C. E. (2020). Indigenous knowledge and disaster risk reduction: An approach to sustainable community development. *International Journal of Disaster Risk Reduction*, 46, 101618. https://doi.org/10.1016/j.ijdrr.2020.101618

Rahman, M.M., Tasnim, F., Uddin, A., Chayan, M.S.I., Arif, M.S.I., & Asikunnaby, H. M. T. (2023). Assessing vulnerability in ethnic Munda community: A study on a cyclone-prone area of Bangladesh. *International Journal of Disaster Risk Reduction*, 95, 103884.

Rasul, G. (2007). The role of Indigenous knowledge in community-based natural resource management: Experiences from Bangladesh. *International Journal of Sustainable Development & World Ecology*, 14(3), 205–217. https://doi.org/10.1080/13504500709469794

Rasul, G., Thapa, G.B., & Zoebisch, M.A. (2004). Determinants of land-use changes in the Chittagong Hill Tracts of Bangladesh. *Applied Geography*, 24(3), 217–240.

Redvers, J. (2020a). "The land is a healer": perspectives on land-based healing from indigenous practitioners in northern Canada. *International Journal of Indigenous Health*, 15(1), 90–109.

Redvers, J. (2020b). "The land is a healer": Perspectives on land-based healing from Indigenous practitioners in northern Canada. *International Journal of Indigenous Health*, 15(1), 90–107.

Revi, A. (2008, April). Climate change risk: An adaptation and mitigation agenda for Indian cities. *Environment and Urbanization*, 20(1), 207–229. https://doi.org/10.1177/0956247808089157

Rokon, S., & Jiggasha, J. (n.d.). (A Collection of Interviews on Climate Change), interviewed Dr. Salimul Huq, pp. 22–27.

Roy, C. R. (2000a). *Land rights of the Indigenous peoples of the Chittagong Hill Tracts, Bangladesh*. IWGIA Document No. 99.

Roy, M. (2000b). Cultural heritage and its role in community-based research: Experiences from Bangladesh. *Bangladesh Journal of Sociology*, 2(1), 29–42.

Roy, R.D. (1998). Land rights: Land use and indigenous peoples in the Chittagong Hill Tracts. In P. Gain (Ed.), *Bangladesh: Land, forest and forest people* (pp. 53–118). Society for Environment and Human Development (SEHD).

Roy, R.D. (2002). *Land and Forest Rights in the Chittagong Hill Tracts*. International Centre for Climate Change and Development (ICIMOD).

Roy, S. (2010). Cultural heritage and community-based research: The Bangladeshi context. In Moody, J. & Small, S. (Ed.), *Research ethics and cultural contexts* (pp. 97–112). Cambridge University Press.

Roy, S. (2024a). Livelihood Resilience of the Indigenous Munda Community in the Bangladesh Sundarbans Forest. In: Leal Filho, W. (eds) *Handbook of Climate Change Resilience*. Springer. https://doi.org/10.1007/978-3-319-93336-8_10

Roy, S. (2024b). *Negotiating Vulnerability and Resilience: Disaster Education, Development, and Children in Coastal Bangladesh*. Indiana University.

Roy, D., & Raja, G. (2000). *The Chittagong Hill Tracts: life and nature at risk*. (No Title).

Saha, K., & Choudhury, S. J. (2021). Traditional settlement and dwelling culture of diasporic: Manipuri community in Bangladesh. *Journal of Business, Society and Science*, 8(1), 93–113.

Saha, K., & Chowdhury, M. S. L. (2021). Traditional Settlement and Dwelling Culture of Diasporic.

Saha, M.K. et al. (2014). Changing Profile of Cyclones in the Context of Climate Change and Adaptation Strategies in Bangladesh Institute of Planners, 2075, 63–78.

Sarkar, O. T., & Mukul, S. A. (2024). Challenges and institutional barriers to forest and landscape restoration in the Chittagong Hill Tracts of Bangladesh. *Land*, 13(4), 558.

Sarkar, J., & Sarkar, S. (2015). Corporate social responsibility in India—An effort to bridge the welfare gap. *Review of Market Integration*, 7(1), 1–36.

Sarwar, A. K. M. G. (2019). Biodiversity and conservation of higher plants in Bangladesh: Present status and future prospects. *Biodiversity and Conservation*, 259–288.

Sarwar, A. K. M. G., Malaker, J. C., & Dutta, M. J. (2008). Floristic composition in the campus of Bangladesh Tea Research Institute – I. Angiosperms. *Journal of Agroforestry and Environment*, 2, 147–152.

Schendel, V. W., Mey, W., & Dewan, K. A. (2001). *The Chittagong Hill Tracts: Living in a borderland*. The University Press Limited.

Seleka, M. L. (2017). Indigenous Women's Leadership and Disaster Resilient Development in Botswana: Fad or Reality?

Shaffril, H. A. M., Idris, K., Sahharon, H., Samah, A. A., & Samah, B. A. (2020). Adaptation towards climate change impacts among highland farmers in Malaysia. *Environmental Science and Pollution Research*, 27, 25209–25219.

Shammin, M.R. et al. (2022). A Framework for Climate Resilient Community-Based Adaptation, In *Climate Change and Community Resilience*, Haque, E. (ed), Springer, pp. 11–30.

Sharmeen, S. (2013). Politics of development and articulation of indigenous identity: The formation of Munda identity in Barind, Bangladesh. *International Journal of Asia Pacific Studies*, 9, 142–160.

Sharmin, Z., & Islam, M. (2013). Consequences of climate change and gender vulnerability: Bangladesh perspective. Available at SSRN 2200116.

Shaw, R., Y. Luo, T.S. Cheong, S. Abdul Halim, S. Chaturvedi, M. Hashizume, G.E. Insarov, Y. Ishikawa, M. Jafari, A. Kitoh, J. Pulhin, C. Singh, K. Vasant, and Z. Zhang, 2022: Asia. In: Climate Change 2022: Impacts, Adaptation and Vulnerability. Contribution of Working Group II to the Sixth Assessment Report of the Intergovernmental Panel on Climate Change [H.-O. Pörtner, D.C. Roberts, M. Tignor, E.S. Poloczanska, K.

Shefer, T., & Bozalek, V. (2022). Wild swimming methodologies for decolonial feminist justice-to-come scholarship. *Feminist Review*, 130(1), 26–43.

Sıddıq, A.B., Habib, A., & Chowdhury, A.H. (2018). Hunting-gathering subsistence of Munda tribe in tropical rainforest of Northeastern Bangladesh. *Cultures and Societies*, 1, 59–71.

Siddique, F., Abbas, R. Z., Mansoor, M. K., Alghamdi, E. S., Saeed, M., Ayaz, M. M., … Hussain, I. (2021). An insight into COVID-19: A 21st century disaster and its relation to immunocompetence and food antioxidants. *Frontiers in Veterinary Science*, 7, 586637.

Simpson, L. B. (2014). Land as pedagogy: Nishnaabeg intelligence and rebellious transformation. Decolonization: Indigeneity, *Education & Society*, 3(3), 1–25.

Simpson, L. B. (2017). *As We Have Always Done: Indigenous Freedom through Radical Resistance*. University of Minnesota Press.

Singh, K. J. (2009). Sacred Groves: The Last Man-Made Carbon Sink. *International Journal of Climate Change: Impacts and Responses*, 1(1), 119–128.

Singh, S. (2020). Yongchak: The Dying Bean in Manipur. https://www.e-pao.net/epSubPageExtractor.asp?src=news_section.opinions.Yongchak_The_Dying_Bean_in_Manipur_By_Sanjenbam_Jugeshwor

Singh, C., Iyer, S., New, M. G., Few, R., Kuchimanchi, B., Segnon, A. C., & Morchain, D. (2021). Interrogating 'effectiveness' in climate change adaptation: 11 guiding principles for adaptation research and practice. *Climate and Development*, 1–15. In Press. https://doi.org/10.1080/17565529.2021.1964937

Singha, R. (2015). Indigenous housing knowledge among the Meetei of Bangladesh [Journal]. *The Journal of Social Studies*, 148, 56–77.

Singha, R. (2022). Retaining Indigenous Knowledge against western technological advancement: A study of Manipuri handloom weavers in Bangladesh Charles Darwin University]. Darwin.

Sioui, M., & R. McLeman (2014). Asserting mino pimàdiziwin on Unceded algonquin territory: Experiences of a Canadian "non-status" first nation in re-establishing its traditional land ethic. *AlterNative*, 10(4), 354–375, doi:10.1177/117718011401000404

SIPCC (2007). IPCC fourth assessment report (AR4). *Climate Change*, 374.

Skogvang (2013). Legal questions regarding mineral exploration and exploitation in indigenous areas. *Michigan State International Law Review*, 22, 321.

Smith, L. (1999a). *Decolonizing Methodologies: Research and Indigenous Peoples*. University of Otago Press.

Smith, L. T. (1999b). *Decolonizing methodologies: Research and indigenous peoples*. Zed Books.

Smith, L. T. (2006). Choosing the margins: The role of research in indigenous struggles for social justice. In N. K. Denzin & M. D. Giardina (Eds.), *Qualitative inquiry and the conservative challenge*. (pp. 151–173). Left Coast Press.

Smith, L. T. (2018). *Decolonizing research: Indigenous storywork as methodology*. Bloomsbury Publishing.

Smith, L. T. (2021). *Decolonizing Methodologies: Research and Indigenous Peoples* (3rd ed.). Zed Books.

Smith, H. A., & Sharp, K. (2012). Indigenous climate knowledges. *Wiley Interdisciplinary Reviews: Climate Change*, 3(5), 467–476.

Steinhauer, E. (2002). Thoughts on an Indigenous research methodology. *Canadian Journal of Native Education*. 26(2), 69–81.

Strong, P. T. (2005). Recent Ethnographic Research on North American Indigenous Peoples, *Annual Review of Anthropology*, 34, 253–268 [CrossRef].

Styres, S. D. (2011). Land as first teacher: A philosophical journeying. *Reflective Practice*, 12(6), 717–731.

Tahir, F., Rasheed, R., Mahmood, S., Chohan, K., & Ahmad, S. R. (2024). REDD+ framework and forest sustainability in Pakistan versus other South Asian countries: A multi-criteria-based analysis. *Environment, Development and Sustainability*, 26(3), 6471–6492.

Taru. (2005). *Disaster Management Plan Blueprint for the Hazira Area Development Committee*; Taru.

Thapa, G. B., & Rasul, G. (2006). Implications of changing national policies on land use in the Chittagong Hill Tracts of Bangladesh. *Journal of Environmental Management*, 81, 441–453. doi: 10.1016/j.jenvman.2005.12.002

The Business Standard (2024). "Vanishing forests, receding waters and impending disaster in the Sangu river basin" https://www.tbsnews.net/features/panorama/vanishing-forests-receding-waters-and-impending-disaster-sangu-river-basin-840501

The Daily Star (2023). Heavy rain inundates parts of Bandarban; 165 mm rainfall recorded in 24hrs. https://www.thedailystar.net/news/bangladesh/news/heavy-rain-inundates-parts-bandarban-165-mm-rainfall-recorded-24hrs-3387881 Retrieved on May 15, 2023.

The Indigenous World 2023. (2023). The Indigenous World, Issue. IWGIA. https://www.iwgia.org/en/resources/indigenous-world

Thomas, R. A. (2005). Honouring the oral traditions of my ancestors through storytelling. Research as Resistance: Critical, Indigenous, and Anti-oppressive Approaches, 237–254.

Thongam, D. (2008). Manipuri nachom. Bangladesh Manipuri Sahitya Parishad.

Topna, E. (2002). *Introducing the Munda people, Platinum Jubilee 1927–2002*, Platinum Jubilee Committee, Dinajpur Catholic Diocese.

Tosun, J. Addressing climate change through climate action. *Clim Action* 1, 1 (2022). https://doi.org/10.1007/s44168-022-00003-8

Tubridy, F. et al. (2022). Managed Retreat and Coastal Climate Change Adaptation: The Environmental Justice Implications and Value of a Coproduction approach. *Land Use Policy*, 114, 105960.

Tuck, E., & Yang, K. W. (2012). Decolonization is not a metaphor. *Decolonization: Indigeneity, Education & Society*, 1(1), 1–40.

Tuck, E., & Yang, K. W. (2014). *Youth Resistance Research and Theories of Change*. Routledge.

Tuck, E. et al. (2015). *Place in Research: Theory, Methodology and Method*. Routledge.

Uddin, M. M. (2014). Land rights of the Indigenous peoples of the Chittagong hill tracts in Bangladesh: A historical overview. *Law Vision*, 13, 20–23.

Ullah, M. S. (2024). Geospatial Modeling of Potential Landslide Hazard Estimation for Better Management in the Bandarban District of Bangladesh. In *Landslide: Susceptibility, Risk Assessment and Sustainability: Application of Geostatistical and Geospatial Modeling* (pp. 669–693). Springer Nature Switzerland.

Ullah, M. N., Rahman, M. M., & Hossain, M. A. (2022). Integrating Indigenous knowledge into environmental management: Lessons from Bangladesh. *Environmental Management*, 69(5), 703–715. https://doi.org/10.1007/s00267-022-01615-0

Van Ryneveld, T.N., & Islar, M. (2022). Coloniality as a Barrier to Climate Action: Hierarchies of Power in a Coal-Based Economy. *Antipode*, 55(3), 958–981.

Van Schendel, W. (2023). Rebuffing Bengali dominance: postcolonial India and Bangladesh. *Critical Asian Studies*, 55(1), 105–135.

Väyrynen, T., Parashar, S., Féron, É., & Confortini, C. C. (Eds.). (2021). *Routledge handbook of feminist peace research*. Routledge.

Wheeler, W. (2001). Thought on the Responsibilities for Indigenous/Native Studies, Indian Studies Department, Saskatchewan Indian Federated College. *The Canadian Journal of Native Studies*, XXI(1), 97–104.

Wigtil, G., Hammer, R. B., Kline, J. D., Mockrin, M. H., Stewart, S. I., Roper, D., & Radeloff, V. C. (2016). Places where wildfire potential and social vulnerability coincide in the coterminous United States. *International Journal of Wildland Fire*, 25(8), 896. https://doi.org/10.1071/wf15109

Wigtil, A. (2022). Is there a place for "plagiarism detection software" in an academic library? *Public Services Quarterly*, 18(2), 129–135.

Wijsman, K., & Feagan, M. (2019). Rethinking knowledge systems for urban resilience: Feminist and decolonial contributions to just transformations. *Environmental Science & Policy*, 98, 70–76.

Wildcat, M. (2020). *Weaving Two Worlds: Indigenous Land-Based Pedagogy and Decolonization*. University of Regina Press.

Wildcat, M., McDonald, M., Irlbacher-Fox, S., & Coulthard, G. (2014). Learning from the Land:

Wilson, S. (2001). What is indigenous research methodology? *Canadian Journal of Native Education*, 25(2), 175–179.

Wilson, S. (2003). Progressing toward an Indigenous research paradigm in Canada and Australia. *Canadian Journal of Native Education*, 27(2), 161–178.

Wilson, S. (2008). *Research is ceremony: Indigenous research methods*. Fernwood Publishing.

Wilson, S. (2020). *Research is ceremony: Indigenous research methods*. Fernwood Publishing.

Wilson, A. C., & Yellow Bird, M. (2005). *For Indigenous Eyes Only: A Decolonization Handbook*. School of American Research Press.

World Bank Group (2024). Climate Risk Country Profile. Bangladesh. Washington, CD. Available at: https://climateknowledgeportal.worldbank.org/sites/default/files/country-profiles/16813-WB_Bangladesh%20Country%20Profile-WEB.pdf

Index

For Product Safety Concerns and Information please contact our EU
representative GPSR@taylorandfrancis.com
Taylor & Francis Verlag GmbH, Kaufingerstraße 24, 80331 München, Germany

www.ingramcontent.com/pod-product-compliance
Lightning Source LLC
LaVergne TN
LVHW021131300926
849160LV00010B/964

* 9 7 8 1 0 3 2 8 8 1 3 4 8 *